THE MAGIC BOOKS

THE
MAGIC
BOOKS

A HISTORY OF
ENCHANTMENT
IN 20 MEDIEVAL
MANUSCRIPTS

ANNE LAWRENCE-MATHERS

YALE UNIVERSITY PRESS
NEW HAVEN AND LONDON

For information about this and other Yale University Press publications, please contact:
U.S. Office: sales.press@yale.edu   yalebooks.com
Europe Office: sales@yaleup.co.uk   yalebooks.co.uk

Set in Adobe Caslon Pro by IDSUK (DataConnection) Ltd
Printed in China

Library of Congress Control Number: 2024943403

ISBN 978-0-300-24443-4

A catalogue record for this book is available from the British Library.

10 9 8 7 6 5 4 3 2 1

CONTENTS

ILLUSTRATIONS

INTRODUCTION

Magic lay at the heart of medieval societies. It attracted the attention of rulers from the Carolingian Empire, through Anglo-Saxon England, the Christian kingdoms of Iberia and Valois France, to Renaissance Italy and Tudor England. Luxury manuscripts, produced for those who held power in both Church and state, attest to the real presence – and practice – of magic in the medieval world's centres of knowledge and government. Astrology, chiromancy, divination, prognostication, ritual magic, angel magic and geomancy are all revealed in these manuscripts' strikingly beautiful, fascinating and strange illuminations. And these manuscripts, many of them now digitised, provide a rich resource for exploring the history of magic in medieval culture.

Focusing on twenty key manuscripts, each newly analysed and explored, this book traces the important developments in magic from the ninth century to the sixteenth. This history ranges across no fewer than three periods of renaissance. At the heart of each was the deliberate revival of selected aspects of classical culture. The ninth-century Carolingian renaissance recreated Roman art, science and literature in order to strengthen the position of the new, self-styled, 'Roman' emperors. The twelfth-century renaissance was more complex, since the enthusiastic pursuit of ancient knowledge embraced updated and expanded forms of that knowledge, which came to western Europe via works in Arabic and Hebrew. The renaissance of the later middle ages blended both approaches. Once again, art, architecture and texts from the classical world were copied and echoed, celebrating the wealth and power of patrons in the process. And once again this exploration of classical culture was expanded to produce something new and fashionable, while benefiting from the glamour of the ancient world. Manuscripts produced in all three periods present magical texts and images that claim sources based on the knowledge of the Greeks and Romans, reinforced by the wisdom of biblical figures, including King Solomon himself.

It follows that claims, real or imagined, of classical origins played an important part in making the forms of magic explored here acceptable – and even prestigious. Magic, like philosophical or scientific texts, was often falsely attributed to classical authors. The fact that the Greeks and the Romans were pagan when much of their knowledge was created was outweighed by the twin facts that both had converted to Christianity and both were presented as the prestigious ancestors of medieval Europeans. By the twelfth century, a growing number of the new states of medieval Europe and of the peoples that inhabited them were claiming descent from such classical heroes as Aeneas. Literary and historical works in new, medieval vernacular languages asserted these claims, while also exploring the magical powers that had allegedly brought heroes such as Alexander and Arthur into being. Chrétien de Troyes, the French poet who did so much to make the adventures of Arthur and his knights fashionable, even stated that medieval culture, building on that of the ancient world, was able to surpass it. As he put it: 'Greece was first in chivalry and learning; then chivalry passed to Rome along with the wealth of transcendent learning that has now come to France.'[1]

All this is important to us for two reasons. The first is that much of the magic to be found in high places – and thus, in the manuscripts explored here – was inherited from the classical world (or else claimed to be so) and benefited from all the prestige that conferred. The second is that this evidence for the importance of magic, in political practice as well as in culture, goes against received opinions, which are still very strongly held. Medieval magic was not confined to the powerless and the poor, any more than it can be simply identified with the later phenomenon of the witch craze. This contradicts the long-held view that belief in magic is inherently irrational, and can only be the product of superstition or ignorance. However, it is in line with a growing body of work on the subject. Most recently, Tabitha Stanmore's study of the provision of magical services in late medieval and early modern England has showed that clients from across the social spectrum paid well for magical healing, love magic and tracking down lost or stolen property.[2] Stanmore also argues convincingly that the attention paid by historians to the witch craze, important as that phenomenon is, has obscured – and even unintentionally distorted – the history of other forms of magic in the same period. Like Stanmore, Jean-Patrice Boudet argues that magic played a major role in medieval societies, to the extent that to omit it is to run the risk of missing a crucial factor in political decision-making.[3] It is well known that both secular law codes and church (canon) law classified many forms of magic as heretical or criminal. For that reason, it is tempting to see magic as a minority or fringe activity, rejected by the educated and the powerful. However, Boudet's survey of magic in medieval courts tends to prove the opposite. The evidence is that the laws reflected the reality, attractions and dangers of magic. In Boudet's words: 'the sovereign powers . . . have not built themselves up only in opposition to magicians, sorcerers and witches but also, to some extent, with their help.'[4]

A crucial element in the building of power by ambitious princes or courtiers was knowledge of likely future trends in politics. Strictly speaking, Christian theology held that precise and accurate knowledge of the future was the preserve of God, and could be made known only by divine revelation and the powers of prophets and saints. However, theologians also held that the Earth and all material bodies living on it were subject to influences received from, or transmitted by, higher bodies and intelligences. The most widely accepted examples of such higher bodies were the planets and the stars. The issue of immaterial intelligences was much more contentious and frequently confined to discussion of angels and their roles.[5] However, for those interested in magic it opened up the possibility that humans, particularly those endowed with deep learning and special skills, could initiate contact with angels or spirits. This was attractive not only in itself, but also for the possibility of obtaining knowledge not available to mere humans. The practices involved in such endeavours, the hopes of those in power to benefit from them and the consequent fears that wicked or gullible humans were falling prey to demonic deceptions run through the chapters of this book.

This is where the focus on specific, important manuscripts plays an important role. The illustrations of significant texts and images, chosen wherever possible from manuscripts that have been fully digitised, show the value placed on the pursuit of hidden knowledge. That value is quite directly visible in the high-quality vellum, the lavish spacing of texts and the luxurious pigments (including precious metals) seen on those folios. Such manuscripts were works of art, often requiring the collaboration of skilled artists and scribes over considerable periods of time. They were also very expensive to acquire, especially in the early middle ages, when there was virtually no such thing as a book trade. Manuscripts were mainly produced in ecclesiastical institutions and changed hands at considerable expense. Those with luxurious illuminations were sufficiently precious to be sent as gifts from one ruler to another.[6] That fact makes it all the more telling that some of the most impressive volumes created under the patronage of Emperor Charlemagne and his immediate successors gave special attention to the celestial bodies. Information on the identities, natures, locations and movements of stars and planets, including knowledge from classical, pagan sources, is proudly displayed. These Carolingian versions of Greek and Roman astronomical and astrological works, examined in Chapter 1, embody the glamour of the ancient world, together with the aspiration to acquire and use its most valuable knowledge.

In a similar way, the manuscripts analysed in Chapter 2, made for important individuals and institutions in Anglo-Saxon England, demonstrate a challenging openness to the illicit practice of divination. Like the Carolingian Empire, late Anglo-Saxon England saw a significant drive to establish Christianity not only as the official religion of the rulers, but also as a fundamental part of cultural and personal life for everyone. That makes the presence of texts for foretelling the future, placed alongside sermons, prayers and other texts for the performance of church services, all the more surprising. Placing a key manuscript in the context of its first, monastic owner and patron, and his dual role in a royal monastery and at the court of the Scandinavian conqueror, King Cnut, sheds light on this puzzling combination of divination and liturgical materials.

Chapter 3, and the manuscripts discussed there, carries the exploration of this dual presence into the period of the twelfth-century renaissance. A prestigious, scholarly and very beautiful manuscript from Canterbury Cathedral is found to contain instructions and tools for divination in the forms of lot-casting and chiromancy (or palm-reading). This manuscript, whose main content is the Book of Psalms, both looks back to a model from the Carolingian Empire and showcases the attractions of new knowledge from the Mediterranean and Islamicate worlds. It was made for Canterbury at a time when a group of internationally renowned scholars, architects, artists and engineers had been assembled by the archbishop. At the same time, Thomas Becket, archdeacon, royal chancellor and future archbishop, was criticised for consulting a diviner and a chiromancer in Canterbury.[7] After this, it is perhaps less surprising that the other manuscript examined in the chapter – a collection of beautifully illuminated divinatory tables – was the work of the well-known monastic chronicler Matthew Paris of St Albans. The expertise in divination demonstrated in this volume may well help to shed new light on the chronicler's access to those in power. It is certain that Matthew was proud of the confidential information, and the personal invitations, given to him by the king himself, and that such favours were highly unusual. Direct engagement with Matthew's book of divinations also reveals the fascinating complexity of the ways in which it provided answers to pressing questions. Like the Anglo-Saxon prayerbook, these later monastic manuscripts place their fortune-telling tools in plain sight – and even present them as contributions to the special guidance which the Church could make available to secular society.

This finding challenges the view that magic was always something outside religious and secular norms. The answer to this seeming contradiction probably lies in the perceived usefulness of magic, as well as in its powers of fascination. Eminent scholars like Albertus Magnus (Albert the Great) and Michael Scot explored both aspects in the thirteenth century. The first was a leading theologian, who played an important part in a group of scholars who developed the theory and practice of natural magic. The second was an expert in astrology, alchemy, chiromancy and physiognomy who provided valued services to both Emperor Frederick II and Pope Honorius III. Chapters 4 to 6 of this book look in depth at this thirteenth-century enthusiasm for learned magic and the benefits it offered. A strange work, tentatively connected to the court of Frederick II, and issued under the pseudonym of Georgius Zotorus Zaparus Fendulus, is the focus of Chapter 5. The treatise is followed from the thirteenth century to the fifteenth, through the medium of four beautiful and tantalising manuscripts. The power of the work and its revelations about the forces embodied in the celestial spheres is demonstrated above all by the high status of the individuals who commissioned and owned these surviving copies. Once again, this was neither the product of popular superstition nor the target of ecclesiastical condemnation.

The books examined in Chapter 4 were created closer to one another in time, and show that the would-be emperor, Alfonso X of Castile, went still further in his pursuit of knowledge and power through magic. He established what might almost be called a school of magic at his court. Here philosophers, astrologers, translators, artists and craftsmen would all be required to collaborate in order to create these beautifully illuminated encyclopaedias of magic. The volumes are also clear about the range of expensive materials and equipment – as well as the rather long periods of time – required to perform the magic itself. The chapter echoes this practical approach by looking at the types of magic and magical rituals set down in the book known as *Picatrix*. It then goes further by examining the surviving illustrations in one of Alfonso's manuscripts, which was acquired by the French royal family in the fourteenth century and finally given to Pope Alexander VIII by Queen Christina of Sweden. These images show the lifelong dedication of a practitioner of ritual magic, harnessing the occult forces of nature and the powers of magical words and names to call upon the angels of the planets.

The magical books created for Alfonso X might appear extremely ambitious in their claims to unite natural and angelic powers in the service of a monarch. However, as the manuscripts examined in Chapter 6 show, the work known as the *Ars notoria* (*Notory Art*) in some ways went even further. It promised full knowledge and understanding of the liberal arts, which were at the heart of medieval knowledge and higher education – and beyond that, even visions of the Virgin and of God Himself. Of all the surviving works of medieval learned magic, this was perhaps the most popular, as the chapter shows. It went through several revisions (one of which was extensive) and it has survived in an unusually large number of copies, which display a range of production values. All versions used esoteric names and words, combined with mysterious visual forms, in complex and long-drawn-out rituals. Analysis of specific copies of the 'prayers' and images, with their accompanying instructions, demonstrates just what was required of those who attempted to acquire the desired revelations and triumphs in this way. A startling autobiographical confession also shows the addictive power of these practices.

After the thirteenth-century high noon of learned magic, it comes as no surprise that the fourteenth century saw the creation of still more extensive and ambitious volumes. These built on existing ideas and techniques, while providing further practical details and instructions. As Chapter 7 shows, the name and reputation of the biblical King Solomon played an important part in many such works. The range of these collections, together with the impression of personal experience and guidance, echoes the rising number of political scandals – in both royal and ecclesiastical courts – in which magic and magicians were involved. Magic has so far appeared as a source of gain and material benefit, but its powers could equally be used to cause harm and destruction. It is no coincidence that magical crimes begin to appear more frequently in legal records, even allowing for the greater quantity of such records which survive.

This rise in both the visibility of, and the reaction against, magic in high places can be compared to the growing concern about witches in the fifteenth century. That, in turn, raises the issue of gender. Magic so far, in the manuscripts examined here, has been almost entirely the preserve of men. However, the growing stereotype of the witch identified women as far more likely to commit the heinous crime of witchcraft. Chapter 8 focuses on one woman who skilfully combined advocacy for the political advantages of astrology with a highly successful literary career. That woman was Christine de Pizan, and a focus on manuscripts directly linked to her, and their contexts in the royal and ducal courts of France, produces intriguing results.

The final chapter examines a form of divination that increased in popularity in the late medieval and early modern periods: geomancy.

Particular attention is given to a beautiful manuscript created for the use of King Richard II of England. This is accompanied by detailed exploration of a Middle English text, available in a modern edition, which explains how to perform geomantic readings.[8] The question of magic and gender arises again in this chapter, since one such treatise appears to have been produced for Eleanor, duchess of Gloucester. The duchess was tried for hiring male magicians and a female witch in connection with the supposedly hoped-for death of Henry VI. The accusations focused on illicit astrology and necromancy, rather than geomancy; but this case is a reminder of both the power and the danger of magic. Nevertheless, the chapter (and indeed the book) ends with a new form of the old theme of the classical origins of magical knowledge. A beautiful manuscript, in the fashionable style of the Italian Renaissance, opens with an idealised image of a geomancer. This authoritative male figure, a philosopher clad in a version of classical robes, rests a hand on a model of the celestial spheres while pondering his book. Once again the glamour of the ancient world provides reassurance that learned enquiry into the forces shaping the future is a goal to be valued, if performed by the right people and in the right way.

1

ASTRONOMY AND ASTROLOGY IN THE CAROLINGIAN EMPIRE

A New View of the Skies

Throughout the medieval period, as in the ancient world, astronomy and astrology were in a very close relationship. In theory, astronomy concerned the basic study of the positions and movements of the stars and planets, while astrology determined the meanings and effects of astronomical data. In practice, however, the terms were often interchangeable, and there was always a considerable overlap between them. This situation was made considerably more complex by the fact that the medieval Church, despite its reputed hostility to all forms of astrology, accepted that the stars and planets did indeed possess God-given powers that could influence earthly matter and creatures, including human beings. The problem was that some applications of astronomical and astrological knowledge went against fundamental Christian teachings on free will, individual responsibility and the natural order established by God. This was especially the case with the form of astrology that cast and interpreted birth horoscopes (whose practitioners were often confusingly called *Mathematici* in Latin). To claim that an individual's destiny was determined by the state of the celestial bodies at the time of that individual's birth went against both free will and responsibility for one's own actions. However, study of astronomical and astrological data to predict coming influences on natural phenomena, like the weather, was both perfectly possible and socially useful.

In practice this meant that the forms of astrology which made predictions concerning individuals raised complex issues. St Augustine, one of the most influential of all the Church Fathers, had been very interested in all forms of astrology in his youth. But when he examined them from a new, Christian, perspective, he perceived that predictions for individuals made human beings into 'slaves of Mars or Venus or all the stars', and concluded that such ideas should be placed 'in the same class as if they were leagues and covenants with devils'.[1] Importantly, Augustine made it clear in the same text that study of the divinely ordained movements of the stars and planets was wholly acceptable. On the basis of the biblical book of Genesis, Augustine also stated firmly that study of celestial positions made possible some very useful predictions, such as weather forecasts, guidance for travellers and the calculation of calendars and dates.[2] No theologian, of course, denied that it was knowledge and study of the stars which had led the pagan magi to Bethlehem and had informed them of the enormous importance of the birth of Christ. However, the brevity of the biblical account, and lack of historical knowledge concerning 'Chaldean' astronomy and astrology, led to considerable ambiguity in theologians' views of the magi and the source of their knowledge.[3]

It is important to note that the personal, or 'judicial', form of astrology was also problematic for makers of secular law. The rulers of the Roman Empire, like rulers before and after them, were concerned about unacceptable attempts to harness the power of the stars. This condemnation was deepened when dissidents were accused of using astrological magic to find out when and how unpopular emperors were going to die.[4] Fear of astrology was increased by the fact that very few Roman scholars were expert in astronomy, geometry and mathematics – in part, at least, because much of the literature on these subjects was in Greek.[5] This gave such knowledge special status and also meant that astrologers could claim possession of scarce and esoteric knowledge.

This was the complex heritage that was passed to the new states of early medieval Europe. They were absorbing both Christianity and versions of classical literature and culture, while continuing to draw important aspects of their sense of identity from their own languages and traditions. The surviving evidence strongly suggests that astronomical knowledge held a high status in this period of transition, and was even part of what made Christianity attractive to the states that succeeded the Roman Empire in northern Europe. Indeed, the calendar of the new religion required accurate observation and calculation of the movements of the heavens. A key textbook on this was written as early as c.700 CE. This was *On Times*, an early work by the Venerable Bede, monk and scholar at Jarrow in the Anglo-Saxon kingdom of Bernicia (later part of Northumbria). In it, Bede explained that a special, nineteen-year cycle is needed to keep the solar and lunar data synchronised in the manner required for calculating the Easter full moon.[6] Perhaps more impressive to secular society was the scientific explanation, and even prediction, of eclipses, which Bede also expounded.[7] However, this in itself did nothing to solve the problem of where astronomy and useful astrology crossed the border into dangerous astrology.

There was, of course, another factor that made knowledge inherited from the ancient world problematic: the issue that it was inherently pagan. This was particularly challenging in the case of study of the celestial bodies, since the planets and the stars not only bore the names of pagan deities, but were sometimes identified with them. For Christians, such ideas were foolish at best and idolatrous at worst, even though the theory that the celestial bodies possessed powers over earthly phenomena was accepted as scientific fact. How, then, could valuable, scientific knowledge be separated from pagan error? This was all the more important, since the Church viewed pagan deities as very likely to be demons, wielding dangerous powers in order to deceive and tempt humans. It was this identification of forbidden astrology with demonic deception and divination that linked it to the emerging definitions of magic. The problem deepened as astronomical and astrological knowledge were updated and expanded by the translation and importation of works originally composed in Arabic. Astrology will be a recurring theme in this book.

In this chapter we focus on the eighth and ninth centuries, and on manuscripts produced in the revived Roman Empire of the early middle ages, the Carolingian Empire. The conscious revival of Roman culture, coupled with a strong emphasis on the spread and consolidation of Christianity, made the renewal and adaptation of ancient astronomical and astrological knowledge an important – though potentially dangerous – matter. Perhaps for that very reason, both Emperor Charlemagne (d. 814) and his successors showed considerable interest in the subject. Surviving manuscripts linked to the imperial court and to powerful ecclesiastical institutions show the effort that went into creating beautiful books which brought together pagan knowledge and Christian applications. Charlemagne himself was ruler of the Franks from 768. Through a series of military campaigns and victories, he established power in northern Italy and central Europe, before being crowned 'Emperor of the Romans' by Pope Leo III in the year 800. His new empire required an impressive and unifying view of history and religion, as well as the convincing adoption and reuse of Roman knowledge and culture. This need drove both scholarship and artistic production in Charlemagne's own reign and that of his son and successor, Louis the Pious (r. 814–40). It was during the latter's reign that strikingly beautiful, richly illuminated manuscripts were produced, containing not the customary religious texts, but rather selections from Greek and Roman astronomy and even astrology. Their importance is demonstrated by the fact that copies were distributed across the empire, as part of the project to create a shared identity linking the Roman past to the post-Roman, Christian present.[8]

At their heart was a text known as the *Phaenomena*. This long description of the known universe, and of collected signs of coming weather, was composed by Aratus, a poet working at the court of

Macedon c.276 BCE, and its title is Greek.[9] Latin translations were made by figures as celebrated as Cicero (the Roman statesman and orator who was much admired in medieval Europe) and Germanicus (who was identified with the emperor of the same name). These are now usually grouped together under a version of the original author's name and known as the *Aratea*. They were highly regarded in the period from the ninth century to the tenth, and again in the Renaissance.[10] Latin versions of the *Phaenomena* or *Aratea* were central to ambitious and lavishly decorated manuscripts. These blended pagan legends, astronomical knowledge and the Christian concept of the universe into a complex whole. The tensions involved in recreating pagan, classical knowledge for the court of a Christian emperor, and the decisions concerning astrology in particular, can be seen through a detailed study of the books.

One type of manuscript is more obviously Christian: it contains a wide-ranging collection of texts and materials focused on the calendar of the Christian Church and the technical processes involved in calculating its central feast of Easter. The link to pagan astronomy is found in the fact that this latter calculation involved astronomically based factors such as the date of the spring equinox and the correlation between full moons, the days of the week and the Roman, solar months of March and April. Central to this field of expertise, known as *computus*, were the apparently simple dating requirements for Easter. These were: that it should always fall on a Sunday; and that it should fall at, or shortly after, the full moon which itself followed the spring equinox. This was why study of both the lunar and the solar calendar was important, and why it was necessary for the Church to adopt a unified method for their correlation and for choosing the appropriate Sunday for Easter.[11] At a time when the emperors of the new Carolingian dynasty were linking territorial conquest with forcible religious conversion, it was all the more important to be able to show that Christian scholars could provide superior answers to questions of both religion and science.

This made it important to assimilate, edit and apply all the resources of classical science, and to be able to answer questions on cosmology, as well as theology. Such an ambition, together with the concept of the new, Frankish emperors as successors to the prestige of Rome, is embodied in these complex assemblages and editions of texts.

One of the most important surviving examples, made for Charlemagne's illegitimate son, Drogo (bishop of Metz), is now in Madrid as MS 3307 in the Biblioteca Nacional. This manuscript is illustrated with beautiful depictions of the constellations as classical deities, figures and objects, which are striking for the absence of any attempt to show the presence of actual stars. It has been analysed in detail, with helpful comments on the complexities of *computus*.[12]

The other version of this material is represented by two manuscripts which, at first sight, are entirely pagan in content and appearance, and are therefore still more important in relation to the problem of astrology. Their main texts are Latin translations of Aratus' long poem and give a complete account of the ancient view of the cosmos. The spherical structure of the universe is expounded, with the sphere of the Earth at its centre. A long section of the poem then describes the stars, their groupings into constellations, and some of the myths associated with them. Their seasonal periods of visibility in the night sky are also covered. There is an account given of the 'wandering stars' or planets, with comments on their movements and significance. The second part of the poem provides a detailed account of various signs of coming weather that may be read in the sky, and offers some information on the lunar months and how they can be correlated with the solar months.

There are manuscripts of two different Latin versions of this work, both linked to the court of Louis the Pious: Leiden, Universiteitsbibliotheek, MS VLQ 79; and London, British Library, Harley MS 647.[13] The images in the Leiden manuscript are designed to look like those created for Roman emperors and courtiers, as can be seen in this depiction on folio 22v.[14] The painting can be compared,

Figure 1.1. The constellation Auriga (usually described
as a charioteer). The goat and two kids were picked out in
the text as giving signs of coming storms. Leiden,
University Libraries, MS VLQ 79, f22v.

for instance, to illustrations in one of the few surviving late Roman manuscripts, a copy of Virgil's *Aeneid* now in the Vatican Library, Rome. The Harley manuscript's images are also very 'Roman' in style but still more complex in their execution, as can be seen in the following depiction of the constellation Pisces on folio 3v.

As this image shows, the manuscript's depictions of constellations are not only classical in style, but also mythical in content. This would have been problematic at the time the manuscript was created. It raises questions about the role of the classical pagan deities in these new, supposedly Christian, depictions of celestial phenomena. It should not be forgotten that, following the

Figure 1.2. The constellation Pisces. London, British Library,
Harley MS 647, f3v.

influential arguments of St Augustine, the majority of church
leaders took the view that pagan deities were not deities at all. The
sympathetic explanation was that they were human beings who had
lived in the early ages of the world and who had been mistakenly
perceived as deities by later, deluded humans; the dominant explanation was that they were deceptive demons and were particularly

dangerous, since they could engage in direct communication with wicked or foolish individuals, and could use their superior powers to make false revelations and false prophecies.[15] In the case of astronomical knowledge there was an additional danger, since both classical and Germanic pagan religions included the belief that the stars and planets provided signs of impending events and even had the power to determine the fate of individuals.[16] The latter idea had political as well as religious implications. Consequently, the hostility of the Roman emperors towards astrology was taken up, if for different reasons, by the ecclesiastical and secular rulers of new, Christian kingdoms in Europe.

However, leaders of the new religion accepted that astronomy, if stripped of pagan and astrological pitfalls, was a valuable field of knowledge. The first book of the Bible, Genesis, established that, on the fourth day of the creation, God had set the great luminaries and the stars in the firmament. They were to give light to the Earth, but were also to be signs and were to have a role in the seasons and the measurements of time. Other biblical books, especially the Book of Job, emphasised that the universe is regulated by divinely established rules, in such a way that the movements of specific stars are linked both to the seasons of the year and to meteorological and agricultural cycles. The biblical account of creation, and the intricacies of the system put in place by God, led to an impressive number of commentaries by leading theologians, and by Augustine in particular. In Book Three of his work *On the Literal Meaning of Genesis*, for instance, Augustine deployed information taken from classical astronomy.[17] What was required was to remove all pagan, heretical and potentially demonic elements. To judge how much of this was done in the imperial manuscripts under discussion, it is necessary to establish some background.

Arguments on this subject could and did become technical and complex; but a useful summary has been provided by St Isidore of Seville (c.560–636). Isidore's encyclopaedic survey of the heritage of classical and early Christian knowledge, the *Etymologies*, was highly influential throughout the early middle ages, and most monasteries appear to have had a copy in their libraries. He defined astronomy as 'the study that contemplates the course of the heavenly bodies and all the figures and positions of the stars'.[18] Isidore's account suggests that astronomy and astrology were intermeshed when first discovered and taught, and he is definite that the greatest astronomer of all was Ptolemy of Alexandria (Book Three, Chapter 26). What he omits to mention is that Ptolemy was also famous as a great astrologer. Isidore had no objection to the study of the courses of the major planets and of the positions of the stars at a chosen season or date. Certain things were unacceptable, however. These were: the attempt to find auguries of the future in the stars; the association of the twelve signs of the zodiac with parts of the human soul or body; and the making of predictions about the character or future of individuals from the stars (Chapter 27). He condemned all such activities as superstitious and against true religion.

In Chapter 71 of Book Three, Isidore goes on to discuss the names allotted to the stars, constellations and planets, and how those names were given.[19] This account condemns those who believe that the pagan deities, mythical figures or supernatural creatures that give their names to heavenly bodies are in any sense real and powerful. Yet Isidore accepts the ongoing use of the names and, by implication, the figures or images associated with them. He then pauses to say that comets actually do signify coming plague, famine or war (although he omits to say that they were also taken to signify the death of kings). He takes the trouble to give an account of the twelve constellations accepted by the pagans as 'signs' (the signs of the zodiac). In Chapter 71, section 22, he describes the make-up of the belt of the zodiac, the signs found on it and their order. He also briefly describes the Sun's annual movement through the zodiac and its resulting relationship to the seasons, while condemning the pagan view that the deities and creatures associated with the signs are real. Perhaps wisely, Isidore does not pin down the exact meaning of the term 'sign' when applied to the zodiac.

Further information on astrology and its practitioners is provided in Book Eight, where Isidore discusses religions. Chapter 9, section 21, defines astrologers as those who perform auguries from the stars. Some study the state of the twelve zodiac signs when an individual is born and attempt to predict that person's character, actions and fate from it. Intriguingly, Isidore says that this was acceptable up to and including the magi who predicted the birth of Christ; but from then on, it has been forbidden. This was a point that required repetition in the centuries which followed.

It is clear that Isidore still thought of both the subject of astronomy and the names allocated to the celestial phenomena themselves in entirely classical terms. He has nothing to say on the religions and beliefs of the pagan peoples who were creating new states in what had been the western Roman Empire. It is therefore easy to be led into thinking that these 'new' pagans had no astronomical traditions of their own and no independent beliefs in the powers of the stars. However, this was not necessarily the case. Moreover, parts of the Carolingian Empire were newly conquered, and in regions such as Saxony the pagan inhabitants were brutally forced to convert to Christianity by Charlemagne. Their view of the heavens in the first half of the ninth century would be in dispute, and the names they gave to celestial phenomena are not well recorded. All this makes the production of luxury copies of works on classical, pagan astronomy appear as a very deliberate decision. Appreciation of such books by the rulers of the new empire emphasised their support for this view of the skies, as well as their identification with the prestige attached to Rome. The assimilation of such an astronomy into Christian scholarship would, however, need careful planning if the pitfalls expounded by Augustine and Isidore were to be avoided.

One miniature in the Leiden manuscript is particularly significant. This image, on folio 93v (Figure 1.3), confronts the viewer

Figure 1.3. 'Horoscope' for the Paschal Moon, 816. Leiden,
University Libraries, MS VLQ 79, f93v.

with the ways in which astrological elements were woven into ways
of thinking about the celestial phenomena and their movements.
The artistic style of the miniature is reminiscent of Roman mosaics
and paintings, as is the use of personifications to represent planets,
months and sectors of the sky. Perhaps most recognisable to modern
viewers are the symbols representing the signs of the zodiac, shown
within the outer red band of the image. Alternating with these are
gold discs, each containing a figure representing a month of the
Roman calendar – although these are not correlated with the months

21

associated with each sign. The large central disc contains a personification of the Earth, around which the seven 'wandering stars' move
on the circles shown in red. Each planet (a category that included
both the Sun and the Moon) is depicted as a personification in a
disc whose size is calibrated to the size of the circle on which it
moves. A striking oddity is that Mercury and Venus, the two planets
closest to the Sun, are each shown as moving (impossibly) on two
quite different circles. The more obvious circles are extraordinary in
that they show the planets orbiting the Sun; but two other circles
bearing the names of the two planets, go around the Earth in the
more approved fashion. So far, the information conveyed is astronomical in Isidore's terms. Even the odd orbits of Venus and
Mercury can be explained by another of the textbooks on astronomy
used by the Carolingians. This is a long, allegorical work by Martianus
Capella, a fifth-century Latin writer about whom almost nothing is
known. The title of his work is *The Marriage of Philology and Mercury*
and, despite its lack of obvious Christian content, it was approved
as a repository of classical knowledge. In this work, the figure of
Astronomy gives a long exposition of the heavens, which includes
the assertion that Mercury and Venus circle the Sun.[20]

It is already clear that the astronomical information lying behind
this image was both ambitious and potentially contentious.
However, there is also evidence of meanings which go beyond the
purely astronomical. This is demonstrated by the positions of
the Sun, the Moon and the other planets. The Sun is shown as in
the sign of Aries with the Moon opposite, in the sign of Libra, and
therefore full. The Sun is in Aries in parts of the months of March
and April, and a full Moon (such as the one indicated here) will
occur at or close to Easter. This means that the date chosen to be
depicted in terms of its planetary configuration is close to Easter;
and analysis of the full range of planetary positions suggests that
the occasion was the Easter full moon and the year in question was
816.[21] No rubric or explanation is provided for the miniature, but it

is clear that the Christian feast was central to its construction. Dating Easter in relation to Roman months and to zodiac signs was in fact an established part of the range of calendar material brought together in works on the Christian scientific practice of *computus*. An authority on this subject was the Venerable Bede, whose work on *The Reckoning of Time* was used in the Carolingian Empire. In Chapter 16 of this work, Bede set out the signs for each month, beginning with Aries, and explained that each sign ends in the middle of the month allocated to it. Thus, Aries will end in the middle of April, and the spring equinox will fall while the Sun is in this sign.[22] Such information, with more besides, was given in the Carolingian computistical volumes discussed below.

Thus far, the image is unproblematic; but other layers of information move into astrology. A first point is that, if a line is drawn from the Earth through the centre of each disc and out to the red circle of the zodiac, each planet will be found to be precisely placed at the start or centre of a zodiac sign. Moreover, Aries, the first sign in the astrological sequence of the year, is shown just above the centre-left of the circle, close to the astrologically important position of the Ascendant. In other words, this image comes close to being a horoscope, in the sense that it depicts the positions of the planets in relation to the signs of the zodiac on a specific date. Those wishing to draw conclusions from these positions are also provided with helpful information in the inscriptions written on the outlines of the circles. The signs in which each planet is most powerful are named, along with those in which the planet is closest to and furthest from the Earth. Judgements as to the relative powers, and even the interactions, of the planets in this image are thus made possible. Does this imply that the owner of this manuscript, almost certainly a member of the imperial court, was prepared to go deeper into astrology than Isidore and Augustine would have condoned? To answer this question, it is necessary to see how much astrological knowledge was actually available.

The fullest exposition of the powers of the planets, and how they were modified by the zodiac signs, was given in Ptolemy's astrological textbook, the *Quadripartitum* (i.e. 'work in four parts'). This was not directly known by Carolingian scholars, but the basic information was given (in slightly varying forms) by the accepted authorities on astronomy. The Roman encyclopaedist Pliny, whose wide-ranging work on *Natural History* was available in the Carolingian Empire, gave an account of the movements of the planets.[23] This states that they have two types of 'elevation'. The first is in relation to Earth and is the point at which they are furthest away (apogee); the second is in relation to the zodiac. Thus, while Saturn is at its apogee in Scorpio, it reaches its zodiacal highest point in Libra. For Mars, the equivalent points are Leo and Capricorn.[24] The Leiden diagram includes extracts of this information, written around the orbits of the relevant planets and adding to the astrological content.

An attempt to make a form of Ptolemy's astrological guidance available in Latin was provided by the text known by its Latin title of *Preceptum canonis Ptolemei*. This offered materials based on Ptolemy's model of the universe, and on his tables for calculating the positions of the planets (particularly the Sun and the Moon). Its address to astrologers is shown by the fact that it lists only stars located near to the belt of the zodiac.[25] The 'worked examples' in the text give dates of 528 and 534, and guidance on adjusting from Ptolemy's location of Alexandria to that of Rome. This suggests that it was produced in Rome shortly after the monk Dionysius Exiguus was commissioned by Pope John I to draw up a fundamental work on *computus* and the calculation of Easter. A copy might therefore have been acquired from Rome. However, no Carolingian copies of the *Preceptum* are known, so direct knowledge of this text is hard to prove. Another possible source is the work of the Roman scholar and monastic pioneer Cassiodorus, who knew a version of Ptolemy's astronomy, as is shown by the relevant section of his *Institutes*.

Cassiodorus' influential work gave definitions of the technical astronomical/astrological phenomena of the 'stations' and retrograde motions of the planets. However, his was a minority view.[26]

This silence may be because, as well as giving instructions for computing the positions of the Sun and the Moon, and for calculating eclipses, the *Preceptum* gave instructions for calculating the degree of the zodiac on the horizon at any given time. This concept is discernible in the Leiden diagram, and yet it was fundamental to the 'superstitious' forms of astrology. Indeed, much of the *Preceptum* would have been problematic for churchmen like Gregory, the late-sixth-century bishop of Tours. As the descendant of a Romanised family, Gregory knew Roman astronomy and applied it in his work *On the Course of the Stars* (*De cursu stellarum*).[27] However, while Gregory saw the traditional, classical star-groupings in the sky, he treated them in a 'modernised' Christian fashion. He gave new, Christian names to some. For instance, the star-groupings of Delphinius and Lyra are, for him, Alpha and Omega, hanging under the arms of the Cross, as in depictions of the Crucifixion of Christ. In other cases, such as his newly invented constellation Quinio, he even made fresh combinations of stars from more than one classical grouping.[28]

Gregory's example makes it all the more interesting that there are, in fact, traces of Ptolemy's works, and their illustrations, in the Carolingian astronomical manuscripts discussed here. To prove this, it is necessary to go back to the two classicising astronomical manuscripts and to look also at a more obviously scholarly manuscript. The latter is now Basel, Universitätsbibliothek, MS AN IV 18, and was probably written and illustrated c.820–35 in the Carolingian monastery of Fulda for the scholar and churchman Hrabanus Maurus. It contains the Latin translation of Aratus' *Phaenomena* made by a writer identified in the early middle ages as Emperor Germanicus. This is also the version used in the more artistic Leiden manuscript. However, the volume from Fulda incorporates additional information

Figure 1.4. The constellation Virgo. Basel,
Universitätsbibliothek, MS AN IV 18, f18v.

on both the pagan myths associated with the constellations and the
positions of the stars within each grouping and figure.[29] It is the
earliest surviving copy of this expanded 'edition' of Aratus' work.[30] Its
contents testify both to the extent of astronomical research undertaken by Carolingian scholars and to the high status accorded to
esoteric knowledge of the celestial bodies – so long as it was associated with the Roman Empire.

Folio 18v of the manuscript, shown here, describes the constellation Virgo and gives information on the placing of the nineteen stars
which are stated to comprise the constellation. The drawing is far
less 'classical' in style than the images in the other two manuscripts;
but the text is closely related, while offering additional information.
This further information itself seems to have been first brought

together and added to Aratus' poem in Alexandria, a little after the time of Ptolemy, and in Greek. This Greek 'edition' appears to have reached the Frankish kingdom in the eighth century, and to have been rather clumsily translated into Latin. That early attempt was revised and improved in the later eighth century, and it is this version which was probably used in the important ninth-century monastic house at Fulda.[31] The effort put into acquiring and transmitting this knowledge is testimony to the importance accorded to it.

To show the impact of this transmission of technical information, it will be useful to compare the two manuscripts. The 'imperial' volume now in Leiden will be considered first. Figure 1.5 shows folio 16v of the manuscript, with a painted miniature representing the constellation Gemini as the Twins or Dioscuri. Rather oddly to the modern eye, they have Christian, gold crosses on top of their white caps, which also show rough squares of burnished gold to represent twin stars. Both wear rust-coloured cloaks, although only that of the figure on the right is fastened – emphasising the three gold stars placed on it. The left figure holds a spear and a club, neither bearing any stars, while the right figure holds a rectangular lyre. Through it, his left hand and wrist may be seen, both with gold stars. In all, the left figure displays twelve such stars and the right figure thirteen, giving a total of twenty-five for the constellation. This exceeds the information given in the accompanying text, which is set out on folio 17r, and so cannot have been based on that alone.

The text is written entirely in capital letters, with rather narrow spaces between words, deliberately imitating the appearance of late Roman luxury manuscripts, like the copy of Virgil's *Aeneid* made at the end of the fourth century and now in the Vatican Library (MS Vat. lat. 3225). Folio 17r has only four short lines of verse, followed by another four on folio 17v. The opening words come from Germanicus' Latin version of Aratus' poem and simply state that the constellation is placed under the head of another star-grouping (in

Figure 1.5. The constellation Gemini. Leiden,
University Libraries, MS VLQ 79, f16v.

fact Cancer, but not named here). The remainder of the text is taken
from yet another Latin version of Aratus, that of Avienius. This
gives more information on the twins themselves, explaining how
Jupiter transformed them into stars, which would shine forever in
the sky. However, even this says nothing about the numbers or placings of actual stars. It is also noteworthy that neither the weapons
held by one twin nor the lyre held by the other are mentioned.

A partial solution to the mystery of the unaccounted-for stars
is that the majority are depicted in accordance with a Latin text
known as the *Epitome Catasterorum*. This work gives brief information on the relevant myths, as well as descriptions of the locations
of stars within the relevant constellation figures.[32] However, even

this allocates only nine stars to the twin closest to Cancer and ten to the other. The designer of the images clearly used yet another source – and it is important to realise that the only way of explaining three of the stars shown is if the star catalogue drawn up by Ptolemy of Alexandria himself was consulted. Sadly, the loss of earlier manuscripts means that it is not known when a Latin version of Ptolemy's star catalogue was produced. Nevertheless, if it is accepted that the images in this manuscript were intended to be impressive and authoritative, then they give still further evidence of Carolingian research into astronomical and astrological sources. That conclusion would be hard to believe without the evidence of the Basel/Fulda manuscript.

What is much better recorded is that direct study of the skies, informed by the reading of Roman texts, was undertaken at the courts of Charlemagne and Louis the Pious. This is described by both their biographers and surviving letters. A product of the work done for Charlemagne was the text known as *De ordine ac positione stellarum in signis* (*On the Order and Position of Stars in Signs*). This text was approved and widely distributed; it omits all mythological tales. Charlemagne also encouraged his senior scholar, Alcuin of York, to teach astronomy to others, as he had already taught the ruler himself. Among the emperor's treasures, listed in his will, was a large table of solid silver which displayed a map of the universe 'in three concentric circles'. Two others showed maps of Rome and Constantinople.[33] Like the manuscripts discussed above, these silver tables embody the strong link between imperial power on the Roman model and mastery of the Roman view of the skies. Sadly, as with the manuscripts, there is no direct record of how these tables were used. However, the letters and biographies demonstrate a powerful desire to read the signs made available in the skies by the movements of the celestial bodies.

Charlemagne's exchange of letters with Alcuin includes two (letters 149 and 155) sent by Alcuin while Charlemagne was fighting the Saxons. These seek to explain the possibly ominous disappearance of Mars from the sky, which lasted from July 797 to July 798, and to account for the phenomenon of its retrograde (or reverse) motion.[34] Since Mars was linked to warfare, its unusual behaviour while an important war was under way would be something on which reassurance was needed. Moreover, Mars is a planet and not a part of a constellation of 'fixed' stars, and thus needed further astronomical research on the part of scholars. Clarification of whether the planets could in fact affect events on Earth was also needed. Charlemagne's own edict of 789, the *Admonitio Generalis*, had condemned all magicians to death, so presumably experts in astronomy and acceptable astrology did not fall into this category. The edict cites the decrees issued by early church councils, and in chapters 16, 18 and 65 declares that unacceptable forms of magic are: the use of strange words and names of strange angels; the making of magical images; giving worship and offerings to pagan deities; and the making of poisonous concoctions. Chapter 65 forbids the making or consultation of auguries, as well as superstitious practices at trees, stones or springs – but there is no mention of superstitious consultation of the skies.[35] The focus here seems to be on lingering Germanic paganism, rather than on 'classical' astrology; but superstitious astrology would count as consultation of auguries. Alcuin does not suggest that Charlemagne was falling into such an error, but emphasises that the planets follow the courses allotted to them by God.[36] Study of their movements, however complex, is thus appropriate for Christians. If Charlemagne, or his army, had been worried by the strange behaviour of Mars, then Alcuin's exposition of planetary movements was designed both to stress that Mars could neither cause nor predict the outcome of the war and to show that this view was based on expert knowledge. Yet the very need to settle the question is significant.

Further evidence as to the range of materials available to Carolingian patrons and the choices they made comes from

Figure 1.6. The constellation Gemini.
Basel, Universitätsbibliothek, MS AN IV 18, f20r.

comparing the 'imperial' volume in Leiden with the 'scholarly' manuscript in Basel. Analysis of the miniature (Figure 1.6) shows similarities to, and differences from, that in the Leiden picture book. Perhaps most significantly of all, the depictions of the stars differ from those in the Leiden version. This image has fewer stars, and the positions of all the chosen stars are related to the information provided in the text. There is no equivalent to the 'Easter horoscope' miniature. It would therefore appear that the original owner of the manuscript now in Leiden was less concerned with astronomical accuracy and more willing to patronise the casting of a horoscope – albeit one with a very Christian focus.

This suggestion of patrons making informed choices is strengthened by the surviving folios of the manuscript now in the British

Library (as Harley MS 647). This has lost several folios and has undergone a series of corrections and additions, both in Frankia and in pre- and post-Conquest England. That complex history is testimony to the high status it enjoyed.[37] Its main section, as the illustration above (Figure 1.2) showed, is written and painted in a style just as classical as that of the Leiden manuscript. It is likely to have been produced for an equally powerful patron, although perhaps slightly later. Its main text is Cicero's Latin verse translation of Aratus, accompanied by long extracts from another 'classical' work. This is the outline of ancient astronomy attributed to Hyginus, a scholar and teacher who lived in Rome in the early first century. It gives versions of the Greek myths related to the constellations.

In the Harley manuscript, however, this mythological material is presented in a very unusual way. This can be seen both in the image of Pisces (Figure 1.2 above), and in the presentation of Sagittarius on folio 6r (Figure 1.7). The name of the constellation is written in clear capitals at the bottom of the folio, while Cicero's poetic lines are set out above it. The text is now badly rubbed, but enough can be seen to show that the lines are written in the distinctive, 'modern' style of script created for Charlemagne and known as Caroline minuscule. Uniquely, the mythological narrative and description from Hyginus are fitted into the body of the centaur that is the 'figure' of the constellation. This is something of a tour de force, and is equally striking for its use of Roman, rustic capitals. Its legibility is preserved by the placement of the dots representing stars around the edges of the figure, despite the astronomical inaccuracy involved. A total of sixteen stars is shown, matching the note between the hind leg and tail of the centaur. Two other manuscripts are known which treat this compound text in the same way; but they are both later than the Harley manuscript.[38]

Perhaps most startling, though, is that this Sagittarius is facing in the opposite direction from that on folio 52v in the Leiden *Aratea* (Figure 1.8). This is, in fact, not an isolated phenomenon.

Figure 1.7. The constellation Sagittarius. London, British Library, Harley MS 647, f6r.

Comparison of the constellation images in these manuscripts both with one another and with the textual descriptions shows that such reversals of constellation figures are a frequent occurrence. The explanation provides still further evidence of the range of

33

Figure 1.8. The constellation Sagittarius.
Leiden, University Libraries, MS VLQ 79, f52v.

knowledge of the heavens available to Carolingian researchers. It begins with Greek astronomy and a difference of opinion as to how constellations should be depicted. A logical and dominant view, attributed to the famous astronomer Hipparchus, was that the figures for each constellation should be shown as seen from Earth. If a reader was to attempt to identify star-groupings in the night sky, this would obviously be helpful, especially if some indication of the relevant placings of individual stars was also given (as happens in the Carolingian manuscripts discussed here). However, both texts and archaeological evidence make it clear that globes showing the positions of the constellation were made and used for the discussion and teaching of astronomy. Some of these globes survived into the medieval period; and Roman mosaics preserved images of others.[39] Such globes created a problem, since they gave the impression that the viewer was placed outside the sphere carrying the stars in their passage around the sky. Should the constellations still be represented as if being seen from Earth, or should they be shown as seen by a theoretical observer looking down on them? If the latter, they would clearly need to be reversed in relation to the standard images. The surviving evidence suggests that this problem caused some confusion. Indeed, as in the Carolingian manuscripts, it often happened that images both 'standard' and 'reversed' occurred side by side in the same volume.

This may be because Carolingian scholars had only images of celestial globes, rather than actual globes. Examples are preserved in ninth-century copies of the revised version of the 'Frankish' translation of Aratus' work mentioned earlier. One such manuscript is now St Gallen, Stiftsbibliothek, MS 902, and it contains a detailed drawing of a celestial globe (on p. 81). Prominent in the drawing, in a way in which it was not in Aratus' poem, is the twelve-part belt of the zodiac, here starting at the left of the globe with the sign of Aries (the Ram). The drawing shows the upper and lower hemispheres of the globe as not aligned with one another, and also muddles some details, suggesting that the artist was unfamiliar with the subject matter. From all this it seems likely that neither artists nor readers could use the images in the luxury astronomical/astrological manuscripts for the purpose of making practical observations and identifications of stars. This problem, together with the still more difficult issue of drawing conclusions as to the significance of such observations, was perhaps best left to experts like Alcuin. Nevertheless, it is clear that careful attention was paid to both the stars and the planets – and not only by scholars, but also by those in power.

An important point is that (as Charlemagne's exchange of letters with Alcuin has already shown) the study of the 'wandering stars' or planets was taken very seriously – even if it brought greater prominence to the problems of superstition and augury-seeking. A practical problem was posed by the fact that locating a given planet was a far more difficult matter than locating a star or constellation, due to the complex movements of the planets. In the absence of Ptolemy's tables for calculating planetary positions, in accordance with his modelling of their orbits, it could be very difficult to be certain where in the sky to look on a given night. For observers with limited technical apparatus at their disposal, there was also the problem that some planets, even when theoretically observable from a given location, might be hard to pick out due to their closeness to the Sun or to other bright stars. This problem was discussed in relation to Mars in the letters of Charlemagne and Alcuin, but it related to other planets as well.[40]

Evidence of knowledge concerning the planets is found in the encyclopaedic volume, now in Madrid, that was mentioned earlier in this chapter. The Madrid manuscript has lost folios at the start and therefore, unfortunately, has no original list of contents. But Book Five includes a text headed *Excerptum de Astrologia* or 'Excerpt on Astrology'. This is, in fact, another version of the familiar list of constellations, with illustrations in an impressive version of the usual classical style. However, the next text is headed *De positione et cursu VII planetarum* or 'On the Position and Orbit of the Seven Planets' (now on folios 63r to 64v). Next come an outline of the courses of the planets and diagrams of their cycles in relation to the zodiac signs. Still further information follows on the movements of the planets within the belt of the zodiac, with futher impressive diagrams, on folios 65v and 66r. Much of this information was carefully excerpted and edited from the account of astronomy given by the Roman administrator and encyclopaedist Pliny, in Book Five of his great *Natural History*.[41]

As this image shows, effort is made in both diagrams to represent the most important, technical information. The left-hand diagram shows the circle of the ecliptic (the apparent course of the Sun

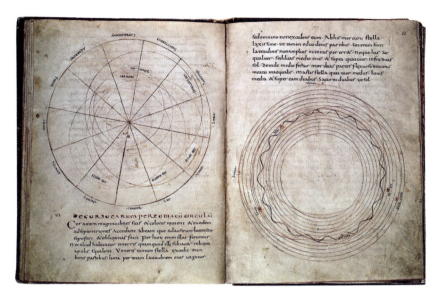

Figure 1.9. Diagrams showing the movements of the planets.
Madrid, Biblioteca Nacional, MS 3307, folios 65v and 66r.

through the year), divided into twelve equal segments for the signs of
the zodiac. Each of these is named. The Earth is in the centre and the
planets are shown to have orbits which are basically circular, varying
in size according to their distance from Earth, but not concentric.
The points at which each reaches its greatest distance from Earth
(apogee) are shown by spiked projections on the appropriate circles,
in the relevant signs. In each case, this projection points to the middle
of the sign, which was usually taken as the exact point of apogee. This
echoes the knowledge deployed in the 'Easter horoscope' in the
Leiden volume. The right-hand diagram is still more technical. This
shows each planet's northward and southward movement within the
belt of the zodiac. However astronomically inaccurate, this diagram
vividly conveys the varying motions and consequent visibility of the
planets, which is discussed in the accompanying text.[42]

Precisely because of its technical and observational difficulties,
such information was highly prized. Cicero had much earlier

expounded the view that astronomy was a clear demonstration of the triumphant power of human intellect and rationality.[43] For Carolingian rulers, it was also a matter of religious and political prestige, as the investment in astronomical and astrological research demonstrates. Further evidence is provided by an exchange of information between Charlemagne and the monk Dungal. This took place in 811, after the death of Alcuin, and seems to have been triggered by a recent embassy to Charlemagne from the Byzantine emperor. The Greek ambassadors appear to have disconcerted the Frankish ruler and his advisers by referring to the fact that there had been no fewer than two solar eclipses in the previous year, although one could not have been seen from Europe. Charlemagne requested that Dungal, a known expert on such subjects, should advise as to whether and how this was possible.

Dungal's reply discusses the issue at length.[44] It shows that he was aware that the emperor and his advisers (like Alcuin) prized the work of Pliny, while Dungal himself drew heavily on yet another source of cosmological knowledge. This was a learned commentary by Macrobius, written in the early fifth century and known as the *Commentary on the Dream of Scipio*. It was very influential in the early middle ages and provided some astronomical information not found in Pliny or Aratus. Dungal expounded the basic structures of the celestial spheres and of the movements of the Sun and Moon, fundamental for understanding the occurrence of eclipses. He concluded that the Greek ambassadors were probably correct. In fact, Dungal seems to have been aware that Byzantine scholars had access to precise planetary data, drawn up by ancient astronomers, that was not available to those in the Latin West. Unsurprisingly, he did not dwell on this, but rather sought to show that the full range of sources available to his own emperor could equal those of the East.

The fact that Macrobius was another accepted authority for Carolingian astronomy and astrology is important because Macrobius gave additional information on the movements of the planets. He went further, believing that it was possible to know the 'starting positions' of the planets and the celestial sphere when first created by God. We learn that, when light first illuminated the Earth and the heavens, Aries was at the midheaven. At the same moment, the Moon was in Cancer, the Sun in Leo, Mercury in Virgo, Venus in Libra, Mars in Scorpio, Jupiter in Sagittarius and Saturn in Capricorn. This information, according to Macrobius, was central to the relationships between the planets and the signs as set out by 'the ancients'.[45] Such explanations were important because they helped to link the 'astrological' information on the powers of planets within different signs to a Christian view of the universe.

If there were any doubt that signs were indeed looked for in the heavens, then a biography of Charlemagne's son and successor, Louis the Pious, removes it. The importance attached to celestial phenomena in this account is such that the author, who remains unidentified, is known to historians as 'the Astronomer'.[46] By this time, a considerable body of observations had been recorded. Annals of the Frankish kingdom had begun to include them in 798. The annal for that year recorded the strange disappearance and reappearance of Mars, about which Charlemagne and Alcuin had corresponded.[47] From then on, both solar and lunar eclipses are noted. The annal for 807 opens with an impressively detailed report, written in a way which shows that the astronomical materials illustrated in this chapter were being put into use. We learn, among other things, that in 806 'there was an eclipse of the Moon on 2 September; at that time the Sun stood in the sixteenth degree of Virgo, the Moon in the sixteenth degree of Pisces'. On 17 March 807, 'the star of Mercury appeared on the Sun as a small black spot, slightly above mid-centre . . . and was visible to us for eight days. But we could not at all notice when it entered and left the Sun, because of the obstructing clouds.'[48] As with the record for 798, so the annal for 810 reflects an astronomical topic that the emperor had checked: it ended with the note that 'in this year both Sun and Moon were eclipsed twice'.[49]

A more detailed glimpse of Louis's own interest in astronomical phenomena is given by the Astronomer, who records that he was summoned to the emperor's private chamber one night in spring 837.[50] The emperor had been tracking the movement of a comet – something which had been accepted as ominous by religious authorities, as noted above. According to the Astronomer, the comet had first appeared in April, in the constellation of Hydra, near Virgo, and had moved through Virgo, Leo, Cancer and Gemini before pausing in Taurus and below Auriga (see Figure 1.1 above). This behaviour troubled the emperor, and his astronomer was summoned for the recorded consultation. It is worth noting that this would not have been recorded if the Astronomer had not also been a biographer, since in this case no letters needed to be exchanged. Louis's own astronomical expertise is flatteringly suggested. The alarming conclusion he drew was that he himself was threatened with death, and the Astronomer responded that he needed more time before giving any formal report. The emperor and his astronomer appear to have been closer observers than monastic scholars, whose observations were less detailed. They are also the only ones to note the odd (and astronomically impossible) halting of this worrying celestial body. Whatever this strange 'star' actually was, what emerges very strongly is that, for both Charlemagne and Louis, knowledge and understanding of the signs provided by the celestial realm were important aspects of imperial power.

The evidence discussed in this chapter makes it possible to come to two conclusions on the knowledge and practice of what might be called magical – or superstitious – astrology in the Carolingian Empire. First, as rulers whose claims to power were linked to their patronage of Roman Christianity, the emperors could not be seen to believe in pagan superstition. Second, the fact that they claimed the status of new Roman emperors created a level of ambiguity when it came to studying and interpreting the stars and planets. The prestige of astronomical and astrological knowledge is clear from the luxurious beauty and scholarly content of the manuscripts analysed here, even though the more technical information regarding the complex movements of planets is presented only in controlled contexts. The very fact that emperors and their advisers engaged in complex and technical consultations as to the meaning of the positions and movements of planets, stars and comets is significant. While paganism and superstition were to be stamped out, interpretations of celestial signs – which, it could be argued, stayed on the right side of the helpfully ambiguous Christian teachings – were obviously valued. It is very striking that two successive emperors engaged in such discussions concerning matters of the highest seriousness – and that manuscripts preserving knowledge collected for the imperial elite were treasured long afterwards.

2

MONKS AND MAGIC IN ANGLO-SAXON ENGLAND

The old-fashioned image of the medieval Church as fighting a constant battle against magic is difficult to apply to the early medieval period. Theologians, preachers and missionaries all understandably paid more attention to the ongoing fight against paganism than to magic. The spread of Christianity across pagan regions of Europe, and its acceptance by local rulers as the official religion of their territories, was a slow process, which lasted from the fifth century until the thirteenth. Indeed, when magic appears in the sources, it is mostly as a subsection of paganism. That is partly a result of the fact that a leading Christian explanation for the apparent powers wielded by pagan deities was that they were in fact demons who had disguised themselves to ensnare gullible humans. Thus, paganism was reduced to an unwitting form of demon worship, while the Church Fathers also taught that the supernatural effects apparently achieved by magicians were in fact the result of collusion with demons. This explanation acknowledged the impressive prophecies and predictions of pagan oracles and magicians, while making it clear that they were the outcome of the knowledge and powers possessed by demons, and thus deadly to Christian souls. As St Augustine pointed out, demons could destroy buildings and cause disease, among other supernatural powers, and thus it was easy for them to predict things which they themselves were about to do. Equally they could see into, and influence, human minds, and then 'predict' the actions of those humans.[1]

The problem for historians is that this overlap between magic and paganism makes it very difficult to tell whether or not magic was consciously being practised in early medieval Europe. Surviving sources are almost exclusively written from the Christian side, which makes the accusations of magical practice on the part of pagans and lapsed Christians difficult to evaluate. For instance, the Anglo-Saxon monk Bede, who wrote his *Ecclesiastical History of the English People* in c.731, records how the recently converted people of Northumbria were struck by plague. In fear, they reverted to practices which Bede labels 'spells, amulets, and other devilish secret arts'.[2] Bede's view is clear, but it is not equally certain that the afflicted people thought they were practising magic.

This contested territory will be explored here. At its centre lies a manuscript produced in Winchester in the early eleventh century (probably before 1029). This is now known as Aelfwine's Prayerbook and is in the British Library, divided into two volumes in the Cotton Collection. These are now Cotton MS Titus D xxvi and D xxvii and have been digitised.[3] The eleventh century may seem surprisingly late for paganism and associated magical practices to be a problem, but this was a time when England was ruled by Scandinavian kings and their supporters, who had only recently – and partially – converted to Christianity.

The portion of the volume that is now Cotton MS Titus D xxvi contains on folio 19v an image depicting a monk – almost certainly Aelfwine – holding a book at the feet of St Peter (holding the keys to heaven). The very fact that this manuscript is classified as a prayerbook demonstrates the central issue for this chapter: can a book be truly Christian if it is also magical? First of all, though, it is important to consider the evidence for the presence of texts of magic in this volume. This chapter will discuss the volume in its original form, as it was before it was split by Cotton. However, all references will be to folios in the current volumes. The texts that might be considered magical are found in a group, which raises immediate questions about their status. They are representatives of a type of text known to historians as 'prognostics', and all have the aim of making it possible for users to produce reliable predictions of future events.[4] It is this matter of claiming to make reliable predictions of uncertain events that raises the question of whether they should be seen as divinatory. Almost all of them involve the observation or use of features in the immediate, physical world, ranging from the day of the week to the phase of the Moon or the occurrence of unusual weather events. Some are linked directly to the individual asking for information and advice, by using the letters of their name or other personal information. To understand the procedures involved, and the forces being invoked, it is essential to look at the texts themselves. These show that the most important of all the factors involved is the Moon, as no fewer than six of the predictive methods focus on that luminary. Of the six, one covers a wide range of possible problems and questions, while the other five are more specialised. Two are predominantly medical, since they relate to the crucial issue of when it was helpful and safe to apply medical treatment in the form of blood-letting. One is intended to provide more general guidance on illness and its likely outcomes, but it could also be classified as useful for those delivering medical treatment and advice. The others are more surprising in a monastic context, since they deal respectively with the interpretation of dreams and the future awaiting newborn children. It is in fact clear that children and their wellbeing were important to those seeking help from this book, since another text is devoted to outlining how children will develop, using the weekday on which they were born as the basis for prediction.

Despite these medical and child-rearing interests, the owner or user of this manuscript was not a doctor. The volume was created for Aelfwine, who was a monk, dean and finally abbot at the powerful New Minster in the royal city of Winchester. Such a man would be able to employ and consult medical professionals, but his own calling was a higher one. Just how high may be suggested by two of the other predictive texts, which are attributed to the biblical prophets Daniel and Esdras (Ezra). Daniel was celebrated for his God-given ability to interpret the foreboding dreams sent to the Old Testament king, Nebuchadnezzar. It is therefore logical that the key to the interpretation of dreams provided in this manuscript is usually attributed to no less a figure than Daniel. The biblical Esdras was also a major prophet, who experienced overpowering visions and was enabled to prophesy events up to the end of the world. His prophetic call was inspired by the angel Uriel, who emphasised how little humans understand of the world created by God, and how dependent they are on divine help and revelation. Interpreting the terrifying wisdom transmitted by the angel to the biblical prophet would not be something to be undertaken lightly, but could be enterprised by a well-trained church leader whose life as a monk placed him in a special relationship with both God and his own society. The importance and complexity of the predictions attributed to Esdras are shown by the fact that two different versions are offered in Aelfwine's book. If this is correct, then these texts should be seen as transmitting revelations granted to biblical prophets, rather than magical divinations.

However, that argument does not apply to the most challenging of the predictive tools offered in this manuscript. This is the text sometimes described as an alphabet prognostic, which is found isolated from the other prognostics, and placed later in the manuscript. It follows a work in Old English, written by Aelfric, the abbot of Eynsham, and intended for the education of monks and priests. Aelfric's text has the Latin title *De temporibus anni* (*On the Times/Seasons of the Year*) and offers instruction on the calendar and its construction, the seasons of the year and their cyclical patterns, and the relation of all this to both human life and the cycles of the stars.[5] A constant theme of the work is that only those inducted into the knowledge offered by the Church can truly understand all this. Aelfric based his work on older texts on the monastic and ecclesiastical learning known as the *computus*. These were written in Latin, the language of the Church and the learned, and a language with which Aelfwine was clearly familiar. The fact that Aelfric's version is in Old English and is very simplified is appropriate for a textbook. In his positions as dean and abbot of the New Minster, Aelfwine would have had some responsibility for the provision of training for monks and priests, and might even have used this text for teaching. This is supported by the fact that Aelfric's work is followed by instructions, also in Old English, on how to calculate the date of Easter, another fundamental skill for clerics.

The alphabet prognostic itself appears on folios 55v, 56r and 56v, filling a space left after the preceding very ecclesiastical material, and was probably added by a different scribe. This raises the question of whether this type of prediction was also seen as appropriate to be taught to priests. Even if that possibility is dismissed, it remains the case that this prognostic was selected for inclusion in a volume which brings together key texts for both the religious and the secular life. It sets out the letters of the Roman alphabet, each on a separate line, and follows each with a sentence giving guidance, prophecy or instruction. No title or heading is given, no author is named, and there are no instructions on how to understand or use it. It is linked to the previous texts only (apparently) by being in Old English. It is worth noting that it is almost certainly a translation from a Latin original, since the Old English alphabet contained extra letters not found here, and also words in Old English did not use the Roman letters 'q' and 'z'. It is also clear that it appears here in a form deemed appropriate for Christian readers and users.

Here is a translation:

A He sets out, and his travels will be safe

B You will find God (or good things) if you make the attempt to do it, and for you it shall be well

C Betokens happiness from the thing in question, but this will not
be easy

D You do not possess the thing you wish for, and you will never
find it

E Bliss will come to you, you will be kept safe always

F This betokens death from death [sic], in this year hope for good
from God

G You will do well to give thought to this

H That will not be granted to you, think of something different

I Begin the thing you want, you will see it through

K Save yourself, to avoid going into danger

L Praise God every day of your life

M God will grant you the power to succeed in what you wish to do

N Stop, your actions will not achieve what needs to be done

O Everyone should praise God with all his might

P If you make the right choice you will not wish to marry

Q Since God is a great and powerful Saviour, you will find bliss

R Reject all sin

S You will be safe if you turn to God, he is great and mighty

T Fear not those who live sinfully and think wickedly

U Bliss will be granted to you, a way to reach peace

X Bliss, guidance, eternal life

Y Betokens peace and security

Z Glory and honour be to you, Lord of Hosts, And glory to the
Father here on earth, In fair and harmonious fellowship With
your own Son and the true Holy Spirit. Amen.

Recent research has found fifty-eight alphabet prognostics surviving
in whole or in part in medieval manuscripts.[6] However, Aelfwine's
version is unusual in several ways, not least in being some 150 years
earlier in date than the next oldest examples. Four have been found
which can be dated to the late twelfth century, and another seven to
the thirteenth century. However, all these are more oracular in the

'meanings' attributed to each letter and more explicit as to their use in dream interpretation. The alphabet in Aelfwine's manuscript is clearly open to being used to provide guidance to those who need it, as well as implicit instructions to trust in God's providence, and yet it has no instructions on how the key letter, which would provide the answer sought by a given individual, is to be identified. Like the prognostics offering advice and insights related to the Moon, this text appears calculated to answer questions of both a spiritual and a worldly nature, but to do so in a way that emphasises the wisdom available to those highly placed in the Church. This, of course, begs the question of whether this wisdom crossed the boundary into what theologians would regard as magical divination.

It is therefore necessary to tackle the problem of definitions, and how far it is true to assert that early medieval monks (and nuns) were deplorably vague on the subject of magic. For this to be correct, it would be necessary for theologians to have paid little attention to the question of magic – or at least for the relevant theological works to be obscure and scarce, so that monastic communities had little opportunity to obtain them. This was simply not the case. The dominant Rule for both monks and nuns from the ninth century in western Europe was that of St Benedict, and this laid down very clear instructions relating to books and reading. Spiritual reading, or *lectio divina*, was a compulsory part of the monastic day, with every monk or nun being formally issued with a book from the community's collection for individual reading and contemplation.[7] This was in addition to the readings during meals and in church services. St Benedict also provided a brief list of the types of text which should be read. Unsurprisingly, all books of the Bible have a prominent place, as do the works of the Church Fathers. Guides to the monastic life also feature, with the works of the early monastic leader Cassian and the Lives of monastic leaders especially emphasised. The Rule goes into no further detail, but one of the most influential of all the Church Fathers was St Augustine, and his works go into considerable detail on the problem of magic. Moreover, all surviving lists of the books available in early medieval European monasteries include works by St Augustine.

What, then, would monastic readers of St Augustine learn about magic? A key passage is found in his work *On Christian Doctrine*.[8] Here Augustine's main target is paganism, which he defines as superstitious, since it involves the making and worshipping of false gods. Still worse, since these false gods are actually demons, worshippers are in fact making pacts with demons. All of the 'magical arts' are included by Augustine in this category of superstitious and demonic pacts. Prominent in his concept of the magical arts are all forms of divination and augury – in other words, attempts to foresee future events. Certain types of supposedly healing practices are also condemned. These are described by Augustine as involving chanting, inscribing mysterious marks or 'characters' and/or tying or hanging up undefined 'things'. For Augustine himself, and his early readers, the pagan and superstitious practices listed here were probably obvious; but problems of translation could arise when these rather sweeping statements were applied in northern and western Europe some five centuries later.

Some further guidance was given in Augustine's short work, *The Divination of Demons*.[9] This was written in response to requests from Christians who were being challenged by pagans who claimed that Greek and Roman knowledge and predictions were superior to Christian ones. Augustine's answer is clear and uncompromising. Pagan gods, as already established, are demons. Demons are fallen angels and thus no longer share any way in divine knowledge, and therefore have no legitimate basis on which to make prophecies. Consequently, all their predictions are traps designed to lure the unwary. Nevertheless, even fallen angels have knowledge, perceptions and powers far exceeding those of humans, and can draw upon these to make convincing predictions and to perform marvels. This is what makes them so dangerous: for these false prophecies and false miracles are the means by which humans are tempted into worshipping demons. The conclusion to be drawn is that the magical arts of divination, augury, magical healing and conjuring are all real – but only because they are performed by demons and by humans who have entered into pacts with demons.

This point was helpfully taken further by Isidore (the seventh-century bishop in Seville), who wrote an enormously influential encyclopaedia (entitled *Aetymologiae* or *Etymologies*).[10] This transmitted a large body of ideas and definitions from the ancient world into Christian Europe, and copies were found in all the great libraries of the early middle ages. Magic as pagan superstition is covered primarily in Book Eight of this twenty-part work, and thus in the section dealing with religion, churches and beliefs. The problems of portents, prognostications and marvels are taken up in more detail in Book Eleven, which deals with human beings and their capacities and experiences.

In Book Eight, magic is reached after the discussions of true and false religions, of poets and of various types of prophets and oracles. Chapter 9 uses the term 'magi' to cover all practitioners of the magic arts, and adds some terms not found in Augustine's definitions. Oracles and necromancy are here added to divination and augury. However, Isidore is just as uncompromising as Augustine in his certainty that all such practices are the result of the machinations of demons. The fact that magic has a history, as well as an ongoing, demonic reality, is emphasised when Isidore offers examples from Greek and Roman history and literature, as well as from the Bible. Still more helpfully, he goes on to set out categories of magical practitioners within the broad terms already given. *Malefici* or workers of evil summon demons and draw upon their powers to disturb the natural world and destroy human wellbeing; they can even kill by invisible means. Almost as terrifying are the *Necromantii* or necromancers, who use offerings of human blood to summon spirits and claim to gain knowledge of future events in this way. Other diviners, using a technique said to have been brought from Persia, gaze into vessels of water and contact demonic spirits in that way. The Roman poet Varro is quoted as an expert on divination, and as the source of the statement that divination can be carried out by use of air, fire and earth, as well as water. However, all forms of divination, even when apparently devout, are to be condemned as drawing upon the powers of demons. This includes the type practised by those astrologers commonly known as *Mathematici* (to be understood as casters of birth horoscopes), who use the stars and planets to make predictions about the lives and fates of individuals. Finally, Augustine's condemnation of magical forms of healing is also echoed by Isidore.

Book Eleven moves on to the subject of human beings, the parts of the human body and the problem of 'portents'. Isidore explains that for pagans these were monstrous beings (human or animal). However, Christians are aware that the rules of nature were established by God, and that nothing can override God's will and power. Thus, portents are part of God's design and appear supernatural only because they exceed normal human experience. The category of portents can be further broken down into signs, omens and prodigies; and all these can, in fact, be sent by God to give warning to human beings. As Isidore says:

Some portents seem to have been created as indications of future events, for God sometimes wants to indicate what is to come through some defects in newborns, and also through dreams and oracles, by which he may foreshadow and indicate future calamity for certain peoples or individuals, as is indeed proved by abundant evidence.

To complicate this issue further, Isidore goes on to state that portents are to be distinguished from both unnatural beings and monstrous peoples, whose appearance does not have any bearing on future events. It thus becomes extremely clear that portents, signs and omens are matters for experts, since the stakes are very high. The unwary interpreter of an apparently supernatural event or creature, or of a powerful dream or a strange light in the sky, could be falling prey to superstitious illusion, and could even be a practitioner of demonic magic.

This is the context within which what can appear as church-sanctioned magic must be understood. Monks, priests and church leaders acknowledged and embraced their responsibilities as leaders of Christian communities; and part of these responsibilities was the duty to guide Christians through the complexities of daily life. Distinguishing genuine portents from demonic divination and superstitious forms of astrology and medicine was a part of this duty, and what has been said above shows that it was no simple matter. The practice of medicine allowed for the use of both natural substances and prayers for divine aid; but anything falling into the categories condemned by Augustine and Isidore was likely to damn the practitioner's (and perhaps the patient's) eternal soul. Moreover, as the Bible demonstrated – and both Augustine and Isidore had made clear – God continued to send messages and warnings to humanity. Thus, some predictions of future events were genuinely to be welcomed as divinely inspired prophecies, even while others were to be condemned as demonically inspired divinations. That being the case, what better repository could there be for the means of making these distinctions and of producing genuine insights into the future than a book of prayers and readings produced for a leader of one of the great monasteries of Anglo-Saxon England? It is therefore the argument advanced here that the apparently supersti-tious or magical elements found in Aelfwine's Prayerbook are in fact the weapons needed to distinguish between acceptable and unacceptable forms of the supernatural.

MAGIC IN ANGLO-SAXON ENGLAND

The next problem is that Aelfwine's manuscript was created in the first half of the eleventh century, and thus some four centuries after

Christianity was first accepted by Anglo-Saxon kingdoms. In the case of Wessex, within which Winchester was located, Bede (Book Three, Chapter 7) records that Cynigils, king of the Gewissae or West Saxons, was baptised in 635 and gave the missionary Birinus the town of Dorchester (on Thames) as a base for further activities.[11] This might suggest that superstitious beliefs and practices of pagan origin were no longer a considered serious issue when Aelfwine was dean of the New Minster in Winchester – probably in the 1020s. Interest in prediction of the future, the meaning of dreams and even the probable characters of newborn babies were apparently things which Aelfwine found entirely acceptable, provided they were done correctly. Nor did his expertise in such things hold back his career, since he became abbot of his prestigious monastery in c.1031 and held that post until his death in 1057. This made Aelfwine a potentially influential figure in the royal court, as well as a leader in the Church. Should the conclusion be that Christian leaders felt so secure in their victory over paganism that they could themselves take on some of the roles of providing insight into the future which had once been performed by pagan diviners and magicians?

This hypothesis might be attractive, but it has a significant flaw. England had experienced major invasions from Scandinavia during the reign of Aethelred II (traditionally known as 'Aethelred the Unready'). The country suffered severely for over thirty years, and eventually its leaders allowed Aethelred to be driven into exile and replaced by the pagan Danish King Swejn in 1013. Swejn died early in 1014 and, despite a final struggle on the part of the Anglo-Saxons, was succeeded by his son Cnut (also known as Canute) after another invasion in 1016. The new king, brought up as a Viking prince, strategically converted to Christianity when he became king in England, and accepted Christian archbishops, bishops and abbots among his advisers. He also married Emma of Normandy, widow of Aethelred, in a Christian ceremony. Indeed, he and Emma acted as joint patrons of the New Minster in Winchester, and Aelfwine seems to have had something of a special relationship with the new rulers. The evidence for this is found in another manuscript in the British Library, Stowe MS 944, which is known as the New Minster 'Book of Life' or *Liber Vitae*.[12]

This manuscript was created in 1031 or 1032, and was mostly written by the same scribe who made Aelfwine's Prayerbook. The 'Book of Life' contains copies of key documents guaranteeing the wealth, spiritual life and political security of the New Minster at this uncertain time. At the head of the list of its monks is the name of Abbot Aelfwine. Most striking of all the unusual features of this manuscript is the set of full-page miniatures with which it opens. The first is on folio 6r and shows Cnut and Emma being accepted as king and queen by Christ and His saints above, and by the monastic community of the New Minster below. They are demonstrating their favour towards the monastery by placing an enormous gold cross on its altar.

Immediately following the royal 'portraits' is another impressive illumination, this one showing an assembly of saints, angels and saintly clerics acclaiming the triumph of God's plan and of Christianity at the Last Judgement. The impression is clearly conveyed that it is the Church that stands between sinful human beings and the power of the Devil. Moreover, the hero in this fight is very much St Peter, who has already appeared in the first image as a patron of Cnut, Emma and the New Minster. Aelfwine is not identified in these images; nevertheless, the suggestion is that the new regime needed the new abbot and his monastery as much as the monks needed royal protection.

Insight into the future would be desperately needed at the time when Aelfwine's Prayerbook was created. This can be dated to the 1020s by the records of the deaths of prominent individuals which were entered into its calendar. Those who died before 1029 were entered by the main scribe when copying the calendar; but those who died from 1029 on were added in different hands. It is also

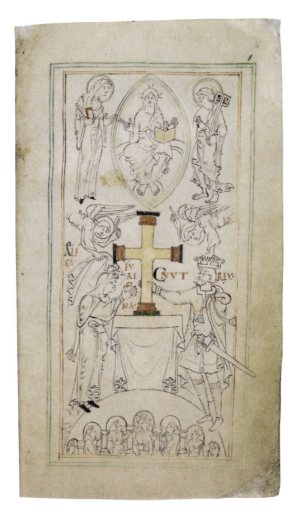

Figure 2.1. Cnut and Emma at the altar of the
New Minster, Winchester. London, British
Library, Stowe MS 944, f6r.

significant that Aelfwine himself has the title of dean here, not
abbot. The volume was thus made during the time when Cnut was
securing his power as king and was eliminating powerful leaders
who posed a threat. Cnut had at first ruled Wessex directly, while
placing newly appointed earls in charge of Northumbria, East

Anglia and Mercia. In the 1020s, he made a fateful decision by giving Earl Godwine (father of the future king, Harold) power over the whole of Wessex itself. It is worth remembering that the majority of the royal estates were in Wessex, as well as much of the royal treasure. Cnut's action suggests that he had great faith in Godwine's loyalty. Nevertheless, having the support of a great monastic house, believed to have been founded by the heroic King Alfred, will also have been of considerable importance in securing the acceptance of his rule in the heartland of the old Anglo-Saxon dynasty. The New Minster had received significant patronage from Alfred's successors, and King Edward had had Alfred's body moved there from the cathedral, giving it strong links to the Wessex dynasty. A monastic leader might well need all possible guidance in attempting to navigate such difficult political waters and in trying to protect those who depended on him.

At this point a new problem arises. The argument so far has suggested that both Aelfwine himself and the new regime taking power in Wessex and Winchester would feel the need for guidance and knowledge as to the future. Does this mean that the 'prognostic' texts found in the Prayerbook should be seen as licensed forms of divination, to be indulged in by monks, despite all that was said above? Such a conclusion would be all the more surprising as Aelfwine and his monks could hardly claim ignorance of the established theological views on divination as demonic illusion. Sadly, the library of the New Minster was burnt down when Winchester was besieged in 1141 during the civil war between King Stephen and the Empress Matilda. Few books seem to have survived the fire, and no catalogue of the monastic library exists. However, the manuscripts discussed in this chapter demonstrate that scribes and artists worked there, and that the Benedictine monastic life was being observed. Moreover, it has long been accepted that both Isidore's *Etymologies* and Augustine's *Christian Doctrine* were well known in Anglo-Saxon England.

It should be noted that kings and their legal officials took a somewhat different view of magic from that of the Church. Legislators from Alfred onwards had emphasised that the use of magic and the employment of magicians were both crimes.[13] For Alfred, the punishment even for receiving magicians into one's home was death. However, while Alfred's grandson, Athelstan, still legislated against magic, its definition as a crime against secular law was much narrower. In Athelstan's law code of c.928, magic is referred to as witchcraft (*wiccecraeftum*). It was illegal in itself, but specific examples were also given: using magical drugs and killing secretly by means of magic. The penalty was automatically death if the victim had died and the accused could not deny the crime; if the alleged perpetrator did deny it, then they were to be subjected to a three-fold ordeal (which was as severe as it sounds). This left it rather unclear what action was to be taken if the alleged victim did not die. The loophole was closed by King Edmund (r. 939–46). His law code specified that anyone who used drugs in magic was to be expelled from Christian society, thus becoming an outlaw – with a life expectancy that would not be high.[14]

Evidence of belief that such magic was actually being practised survives in a document concerning an estate in Northamptonshire acquired in c.970 by Aethelwold, bishop of Winchester. This states that the land had earlier been confiscated from a widow and her son because they had secretly used a 'murderous instrument' to drive metal pins into a neighbour. The two were seized and searched when the neighbour became ill (although he did not die) and were taken to London. The mother was drowned at London Bridge, but the son escaped and became an outlaw. King Edgar then took control of the land, which was ultimately purchased by Aethelwold. The words 'magic' and 'witchcraft' are not used in this document, but the crime fits the legal definitions given above. The wording of the record is very brief and simply states that 'the woman was drowned at London Bridge'. It is not clear whether this was an execution or an unusual form of ordeal; but royal approval of the action is suggested by the fact that the estate was taken over by the king, who awarded it to the alleged victim. It was only at a later stage that the estate came into the possession of the Church, and the document makes no suggestion of clerical involvement in the hunting down of the widow and her son as practitioners of murderous magic.[15] In late-tenth-century England, harmful magic appears to have been very real as both a crime against true religion and a tool for harming and killing other people. When it took the form of deviation from Christian faith, it was a matter for the Church; but when it was directly used to attack other people, it was the concern of the king and his legal officers.

Thus, we have seen that magic was both real and serious for tenth-century kings, as well as for contemporary religious leaders, though at the same time it was given little space in law codes. It became more prominent in the early eleventh century, at the very time when Aelfwine was training as a monk and rising within the New Minster. The reign of Cnut saw the production of numerous texts, both spiritual and legal, which went in detail into the definition, nature and seriousness of magical crimes. It should be noted that most of these were the work of just two powerful individuals. However, it is not convincing to dismiss the texts as merely evidence of eccentricity or obsession on their part. The first of these is Wulfstan, a trusted adviser to Cnut, who became bishop of Worcester and archbishop of York. He appears to have been involved in the drafting of a letter to the English people, in which Cnut attested to his good intentions as their new king. The text survives in a copy of the Gospels which is still in the Chapter Library of York Minster (MS Add. 1). This is a beautiful and richly illuminated manuscript, probably produced at Canterbury Cathedral in c.1000. It was already at York in the early eleventh century, and Archbishop Wulfstan's own handwriting appears in some of the notes and additional texts that were copied into it there. One such text is the 'Letter of Cnut to the English', which starts by setting out the actions which Cnut has already taken as a good king.[16] It goes on to express his awareness of his duties to God and religion, as well as to his people, and to specify that Earl Thorkel is to be his regent in enforcing both royal and church law. A list of the most serious types of criminal is given, and this includes wizards and sorceresses, as well as murderers, perjurers and adulterers. Clearly this gives magic a high ranking as a crime.

Cnut's law code, again showing the influence of Wulfstan, goes to unusual lengths in forbidding 'every heathen practice' and making such practices crimes under secular law. It goes on to specify that the worshipping of heathen gods, or of the Sun and Moon, fire or water, stones, springs or trees are all heathen practices. So, too, are witchcraft, killing by means of sacrificial offerings or divinations, and giving consent to such wicked 'delusions'. A final piece of evidence associated with Wulfstan's work as archbishop of York is the text known as 'The law of the Northumbrian priests'.[17] This survives in an eleventh-century manuscript, which may be either from York or from Worcester (Cambridge, Corpus Christi College, MS 201). It provides striking examples of behaviour and customs which would have shocked a good churchman like Wulfstan, including an apparent tendency of priests to defy the authority of their bishop or to be ignorant of important things such as the timing of baptisms, feasts and fasts, or the correct way in which mass was to be celebrated. Priests also seem to have been engaged in stealing from churches, assisting criminals, living with women, getting into fights and even taking weapons into church. The behaviour of lay people was still worse, since assaults upon priests – and even their murder – have to be forbidden. Cnut's legislation against heathenism and magic is echoed here, but with more detail. It is specified that heathen practice includes divination, as well as sacrifice to or worship of pagan gods, and all forms of witchcraft. A series of heavy fines is set out for these crimes, calibrated to the wealth of the perpetrator. Even if individuals are not caught in the act, the discovery of offerings and other such things at trees, stones or wells is evidence against the landowner. Clearly, the close association between paganism and magic which Augustine had made is being applied here; but it is equally clear that practices identifiable as deliberate magic and witchcraft were believed to be widespread.

The same sense of real urgency in response to what was seen as a growing threat from heathenism and magic appears in the writings of Aelfric. Unlike Wulfstan, Aelfric was not a royal adviser, but he did play an important role in reaffirming the core teachings of Christianity and making them available in formats that could be used to educate lay people. He produced sets of sermons and descriptions of the lives and miracles of saints that were written in Old English (rather than learned Latin) and were clearly intended to evoke emotional reactions, as well as to impart necessary information. Aelfric's own learning is clear, since he drew on theological works and sermons by early leaders of the Church when composing his own works; but the repetition of warnings against heathenism and magic suggests that this was a real issue. In a sermon on the subject of false gods and in another on auguries and divination, Aelfric gives what appear to be contemporary examples, rather than echoes of early sources.[18]

The sermon on auguries first establishes that it is the duty of Christian leaders to protect society from idolatry – which is defined as turning away from Christianity and 'yielding to diabolical heathenism'. Anyone who does this is following the instructions of the Devil, and it can include actions as apparently minor as using magic means in order to get answers to questions or to treat illness. It is not only wrong to perform such acts, but also to consult magicians who perform them. The practice of drawing lots receives a more nuanced treatment, since if it is done to divide up land or possessions, then it is harmless. Attempts to gain insight into future events would clearly fall into the category of auguries and divination, however. Anyone who seeks to gain such knowledge from things like sneezes, or the behaviour of animals, is also indulging in magical practices and turning away from their faith. Looking for guidance in unacceptable places seems to be a prevalent custom, at least in Aelfric's view. People planning to marry, to go on a journey, to engage in brewing or to embark on any enterprise are all prone to do it. So are those with newborn children. Those who attribute special qualities of good or bad luck to the days of the week are also falling into paganism and superstition.

This list may give the impression that Aelfric was simply combating popular customs and practices; but he goes on to talk of the existence of magicians and witches. These appear to be professionals, or at least to offer specialised services, including the treatment of illness. Aelfric's statement that witches are supposed to be able to make true predictions about the future, whereas in fact they are actually being used by the Devil, is too reminiscent of Augustine's teaching to be taken at face value. However, his assertion that it is practitioners of magic who teach that offerings should be made at trees, stones and springs echoes contemporaries, as well as the Church Fathers. His condemnation of women who go to crossroads and pull their infants 'through the earth' is so brief that it suggests his audience would know what he was talking about; however, his list of contraception, abortion and love potions – all equally wrong types of behaviour – is rather more standard.

Resorting to magical rituals in order to acquire knowledge of the future appears to have been especially tempting at New Year, according to Aelfric's sermon for the First of January.[19] He first explains that this day is the eighth since the birth of Christ and thus the day on which the baby Jesus was circumcised. Both this concept and the Roman date for New Year needed careful explanation for Aelfric's audience. The link between the two led Aelfric to a condemnation of the foolish people who indulge in the heathen practice of divination at such times, and thereby abandon their Christianity.

Interestingly, Aelfric here takes pains to demonstrate the superiority of Christian knowledge of the natural world over that of heathens. Thus, foolish pagans allow the phase of the Moon or the day of the week to dominate their decision-making, and fear the existence of unnatural creatures outside God's creation. Christians, however, understand that all living creatures are part of God's plan, and that the Moon has entirely natural powers over waters and tides. It is these natural powers which mean that living bodies, just like the tides, are affected as the Moon waxes and wanes.

This final point is important in understanding the potentially magical elements of Aelfwine's Prayerbook. The condition of England in the 1020s was clearly such that there was a perceived danger of established Christian belief and practice being affected, or even overwhelmed, by pagan ideas and customs. Wulfstan and Aelfric both seem to fear not only outright paganism, but also a serious blurring of the boundary between it and Christianity. Given that Cnut himself, like many of his newly installed aristocrats and officials, was apparently ambiguous about religion, this is highly understandable. Since Cnut's father, Swejn Forkbeard, had accepted Christianity and had married a Christian princess from Poland (Cnut's mother) it is likely that Cnut himself had been baptised before becoming king in England. Adam of Bremen's near-contemporary chronicle asserted that this was the case, and recorded that Cnut's baptismal name was Lambert. Others were less certain, and it is clear that there was no fundamental rejection of the cultural heritage that went with Scandinavian paganism. Poetry in Old Norse, and said to have been composed for Cnut, is recorded in later, Icelandic texts. This poetry makes heavy use of pagan imagery and assumes that the audience will understand. Modern historians (Judith Jesch and Matthew Townend) have paid particular attention to the poems known as the *Knútsdrápur*, and especially the one addressed to Cnut in his capacity as ruler of England.[20] It uses the names of pagan gods when referring to Cnut, and draws upon concepts such as the Valkyries and Iggdrasill (the world tree or 'tree of life'). At the same time, it refers to Cnut as conqueror of England and as having a close relationship with the Christian God, given the titles of 'ruler of monks' and 'lord of everything'. This blurring of boundaries has been described as a knowing use of heathen motifs in a Christian context; the relevance for discussion of Aelfwine and his Prayerbook becomes even stronger when it is considered that Cnut's main court in England is believed to have been at Winchester.

In this situation, rulers who failed to agree to rather technical distinctions between acceptable and unacceptable methods of seeking insight and knowledge would have to be treated with tact. A bishop or abbot who publicly condemned their behaviour might provoke not just retaliation, but a potential confrontation between rival groups within this divided society. Showing that Christianity offered superior tools for forecasting future developments within both the natural world and human society would be a very effective strategy – so long as this superiority could be demonstrated. Placing the tools on offer within the framework of an all-encompassing model of the structure and function of God's creation, and the place of humans within it, would thus be a powerful weapon in the contest against creeping paganism.

The next question has to be whether the Prayerbook and its contents succeed in this challenging task. Can Aelfwine be seen to have provided the counsel and practical guidance that the new rulers of England required, and to have done so within a Christian, rather than a pagan or magical world view? The example of the prophetic alphabet has already been given above. This context also makes the inclusion of prognostics dealing with the futures of newborn children much more explicable, given that Aelfwine, as a prominent member of the royally founded monastery in the Anglo-Saxon capital of Winchester, would be dealing with members of the royal court on a fairly regular basis. The image of the abbey's royal patrons, discussed earlier in this chapter, includes the queen as well as the king, and the choice of a successor (or successors) for Cnut's territories, including England itself, was a very real issue: Cnut had more than one wife, and his son with Emma was not his only heir. Moreover, Emma's sons from her first marriage, to Aethelred (known as the 'Unready'), had claims to England. Privileged insight into the nature and future wellbeing of children was potentially very important in this context. Equally important would be the guidance offered in the choice of positive days on which to undertake activities of varying sorts, from healthcare to travel to warfare; and this was provided by the 'lunary' found on folios 27r to 29v of Aelfwine's book.[21] Both the Bible and medieval chronicles emphasise the importance placed on powerful dreams experienced by kings and rulers, and the text known by its Latin title as *Somniale Danielis* or more simply in English as *Dreambook* also provided this.[22] For instance, if the dream showed a ring being given to the dreamer, this signified safety and security; but a fall into a ditch warned of deception and falsity. The fact that this guidance was placed side by side not only with a wide range of prayers and religious services, but also with clear and confident charts, diagrams and calculations concerning the natural world, the structure of the year and the relationship between natural cycles and the Christian calendar, was also very significant. This was evidence of the superior knowledge of the world, including of human beings and their place in that world, provided by Christianity – or at least it had this status in the eyes of churchmen like Aelfwine. Whether Cnut or Emma ever actually asked for such guidance is not recorded.

As to the question of whether or not the prognostics included in Aelfwine's book should be considered to be magic in the negative senses emphasised by Augustine and Isidore, the conclusion must surely be that they would have fallen into the neutral category later classified as 'natural magic'. This was a subject much debated from the twelfth century onwards, when large amounts of new scientific knowledge concerning the natural world and its marvellous powers were translated from Arabic into Latin.[23] The term itself was not used in Anglo-Saxon England; but it could be argued that Aelfwine's fascinating book provides tantalising evidence that the idea not only existed, but had real political value. A reading of St Augustine's fundamental *City of God* would emphasise that the power of God is 'occult' in the sense of being hidden from human perceptions, and yet is pervasive throughout the natural world.[24] It is this occult power which gives being and form to all that exists, and it is present throughout creation. Since this power is immune to human interference, both it and its effects cannot be altered, even by demons. However, it is possible for humans to perceive the powers and qualities present in natural forms and substances, and to make use of that knowledge. Elsewhere, when discussing the powers of the stars as signs of things to come, Augustine argued strongly against the idea that celestial bodies could predict the future actions of human beings possessed of free will; but he accepted that the stars could be viewed as signs of coming natural changes caused by celestial influences. Thus, the influence of the planets (and especially of the Moon) on earthly bodies was something that fitted unproblematically into this view of natural forces. It is not too much of a stretch to argue that the effects of the planets on earthly phenomena such as the weather, fertility of crops and animals or susceptibility to disease would vary according to the position and strength of the planets. Thus, the prognostics in Aelfwine's book can be seen as a very early example of what would later come to be known as natural magic – and certainly not as examples of what Augustine would have condemned as superstitious ideas and practices. The question of natural magic, and the specific practices which contributed to it, took on greater reality in the twelfth century, as the next chapter shows.

3

—◇—

CLERICAL SIGN-READING AND THE PROBLEM OF DIVINATION

It has already become evident that, while early medieval theologians produced apparently clear definitions of magic and its dangers, applying these in practice was often complex. It would also be naïve to deny that the social status of an individual suspected of magical activities made an enormous difference. However, it also appears that magic was regarded as something closely linked to paganism, rather than being a major problem in its own right. This chapter moves on to a period of great cultural and intellectual change, which affected knowledge and perceptions of magic, along with much else. Volumes have been written about the self-conscious revival of classical knowledge, which played an important part in European scholarship in the twelfth century. However, it is only relatively recently that this picture has been broadened to take account of two further factors. The first is that many of the texts newly available in Latin were translated from Arabic and incorporated new ideas, rather than long-lost old ones. The second is that much of this newly received knowledge can be identified as dealing with astrology, magic and divination. Moreover, at least at first, even rituals and procedures which appear in breach of the theological rules were accepted and transmitted. Such rituals and procedures appear to have been practised quite openly – although by and for the powerful minority. This chapter centres on the evidence provided by two manuscripts: one made for Canterbury Cathedral Priory around the middle of the twelfth century and the other for the abbey of St

Albans in the early thirteenth. Both raise important questions about the readiness of those in power to put perceived advantages before theoretical questions.

The more obviously magical manuscript is the second, which is the work of the well-known chronicler and artist of St Albans, the monk Matthew Paris. Its contents are entirely devoted to providing answers to questions about future events through various forms of divination. The volume provides sets of impressively detailed, and attractively illustrated, tables for both questions and answers, invoking classical and biblical authorities as validation. It was clearly admired, rather than rejected, since high-quality copies were made in later centuries. One such copy survives as Digby MS 46 in the Bodleian Library, Oxford (see Figure 3.12 below). Matthew Paris' original is now Ashmole MS 304 in the Bodleian Library, Oxford, and is a beautiful and complex collection of lot-casting devices, complete with the remains of a rotating dial (or 'volvelle') for generating key numbers as part of the divinatory process. So different is it from anything else produced by a monk who is primarily known as a chronicler and political commentator that until recently it has received relatively little attention.[1]

The first manuscript is different, in that its main content is a complex presentation of the Book of Psalms, with no fewer than three Latin versions of the main text, accompanied by translations and commentaries in three languages (Latin, Anglo-Norman French and Old English). It is sometimes described as a liturgical manuscript, partly because the psalms were an important component of religious services. Its rich illuminations also suggest that it was meant for use in an elevated context. However, it would be of little practical use during a service, and its calendar is strikingly brief and unhelpful in relation to the dating of many major religious feasts. Moreover, its textual contents far exceed what was needed for learning or chanting the psalms. It would be most fitted for scholarly study – except that would not require the complex and striking illuminations which appear on every folio. In short, and like Ashmole 304, this is a manuscript which is as puzzling as it is beautiful.

What makes this psalter important for the history of magic is that it also contains the earliest known Latin text of a set of instructions for palm-reading (then known as chiromancy), together with a simple version of the same divinatory or 'fortune-telling' technique later used in much more elaborate ways by Matthew Paris.[2] Both manuscripts were produced within major monastic communities, and both display very high production standards. Thus, they are linked not only by their monastic context, but also in being luxury manuscripts. Taken together, they provide important information on how magic, in the form of means for answering questions about the future of individuals, was received by both powerful monasteries and their even more powerful patrons.

The psalter from Canterbury Cathedral is now in the library of Trinity College, Cambridge, and has been digitised.[3] It is usually known as the Eadwine Psalter, after its chief scribe, who appears in a portrait on folio 283v, towards the end of the book, in the same section as the texts on chiromancy and divination. Eadwine's exact role in the planning and production of the manuscript is not stated, but this portrait is unique in having a tribute to him, in Latin verse, entered into its border. The poem celebrates Eadwine through an imaginary dialogue with his work:

> SCRIBE: I am the chief of scribes, and neither my praise nor my fame shall die; cry out, oh my letter, who I am.
> LETTER: The fame of your writing foretells that your name, Eadwine, shown in this painted figure, will live through the ages. The beauty of this book shows your genius. Receive, O God, both the book and the scribe as an acceptable gift.[4]

It is clear that Eadwine was no ordinary monk. Moreover, an earlier tribute to him appears at the end of Psalm 150 (on folio 262r) which emphasises his role in the creation of this manuscript. However, there is also overwhelming evidence that this manuscript involved work by other scribes and artists within the priory. It follows that the making of the manuscript, and the costs involved, must have been approved by the prior before work began. Thus, the presence of the divinatory texts cannot be attributed to Eadwine alone. This makes their presence, and the high standard of their presentation, all the more significant.

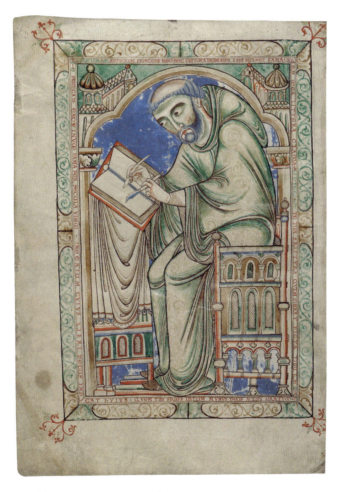

Figure 3.1. Portrait of Eadwine, chief of scribes.
Cambridge, Trinity College, MS R 17 1, f283v.

The first point to be made is that this psalter is linked to Canterbury Cathedral and its priory both by the illustrations provided for each psalm and by the famous, two-page illustration of the new system providing water to all parts of the monastic precincts. The latter follows the portrait of Eadwine, and appears to be a record of something of which the community was very proud. As with the divinatory texts, the relationship of the water works to Eadwine himself is left unstated; but it is suggestive that all were added in the final part of this manuscript, which had already been associated with Eadwine by the portrait. Assuming that the placing was deliberate, the knowledge of the 'prophetic' texts, the work of Eadwine the Scribe and the installation of a complete water system are presented as linked achievements.

The installation of the new water system seems to have taken place in the middle of the twelfth century – at very much the same time as the main phase of work on Eadwine's Psalter. This was a time of major investment in the cathedral priory, under the leadership of Prior Wibert (1152–67).[5] Thomas Becket was also making his mark as a Canterbury archdeacon, before being made chancellor of England in 1155. This promotion meant that Becket spent little time in Canterbury for some years, although that changed when Henry II forced him to take on the role of archbishop of Canterbury. Nevertheless, Becket's friend, the theologian and later bishop of Chartres, John of Salisbury, recorded disapprovingly that it was while he was chancellor that Becket consulted a palm-reader, as well as a seer. This was no light matter, since the consultation was intended to provide guidance in relation to Henry II's campaign against North Wales (in 1157). John reproved Becket for relying on the false predictions of such deceivers, and pointed out that the campaign was disastrous for Becket personally.[6] Nevertheless, John's account of the popularity of various forms of divination at the Anglo-Norman court and among the clergy suggests strongly that such knowledge, and its practitioners, were valued. It is important to remember that this was a time when large quantities of exciting, new works on medicine, astronomy, astrology and the natural world were being produced. Study of this knowledge was highly valued and was actively recommended to rulers by its supporters.

One such was the cleric and itinerant scholar, Adelard of Bath, who dedicated one of his works to a prince who was almost certainly the future Henry II.[7] Adelard had received patronage from John of Tours, who served as bishop of Bath and Wells (1088–1122) after an earlier career as a physician.[8] This career change was not as surprising as it might seem. Baldwin, abbot of Bury St Edmunds, Gilbert Maminot, bishop of Lisieux, and Faricius, abbot of Abingdon, had also provided medical expertise to rulers before being appointed to their positions in the Church, a fact which indicates the value placed on such knowledge.[9] It seems that John's decision to move his episcopal seat from Wells to Bath was influenced by his interest in the famous hot springs and their powers. He restored the Roman baths and attracted doctors to the city, as well as funding a school.[10] Adelard's studies in France, and his decision to go further afield in search of Arabic knowledge, were supported by John.[11] The fruits of these studies were offered to those interested when Adelard returned to England after researching in Sicily and Antioch. They included early translations of planetary tables and of Euclid's *Elements*, as well as treatises on: images of astral magic; the use of the astrolabe; astrology; and the abacus. It was the treatise on the astrolabe that was dedicated to 'Henry, young relative of the king'. There is no mention of either chiromancy or water-engineering in this long list of Adelard's areas of expertise, but his career offers considerable evidence of the demand for exciting knowledge in twelfth-century England.

This does not mean that all forms of magic were regarded with tolerance by leaders of the Church in England. Already in 1125, at a synod held at Westminster under the authority of the papal legate, John of Crema, attention was paid to the condemnation of magical

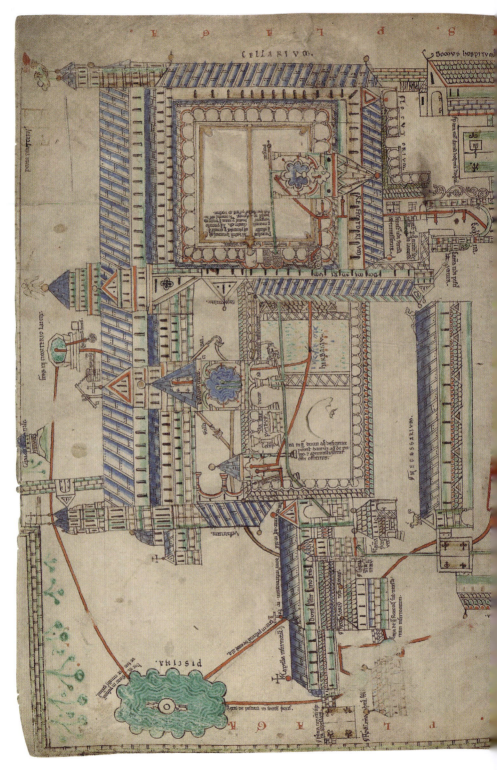

Figure 3.2. The water system at Christ Church, Canterbury. Cambridge, Trinity College MS R 17 1, ff284v and 285r.

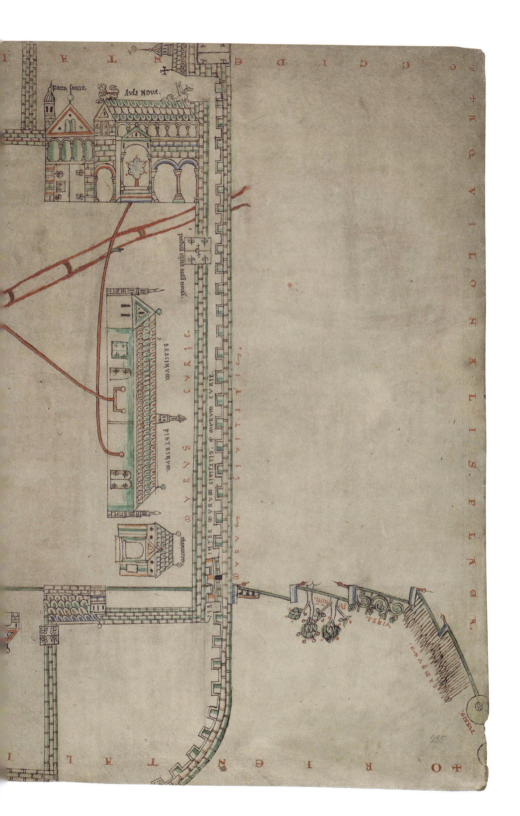

Porta buris.   Aula nova.

porta uzru uult nona.

BASILICVM

PISTRINVM

MVRVS   CIVITATIS

MVRVS   CIVITATIS

MVRVS CIVITATIS

COQVINA

VINEA

CAMPUS

285

attempts to foretell the future. As recorded by the monastic chronicler John of Worcester, himself a student of the new astronomical knowledge, canon 15 of the 17 agreed by the synod stated: 'We decree that sorcerers, fortune tellers, and those who deal in divination of any kind, shall be excommunicated; and we brand them with perpetual infamy.'[12] Whether this suggests that John of Crema was alarmed by what he found in the English churches he visited is not clear, but it would certainly mean that users of texts like the Canterbury one on chiromancy would need to be careful. The patronage of a prestigious client, like Thomas Becket, would provide important protection for a practitioner of this new art, even if it could not avert the disapproval of scholars like John of Salisbury.

It is now important to look at the chiromancy text itself. It is copied formally and clearly, although without the diagram which it seems to need. Readers were perhaps expected to study their own palms for reference. It is introduced by an impressive initial 'L' in bright gold, and the scribe who copied it was one of the team who worked together on the manuscript. He does not appear to have been a trained palm-reader himself, since he misreads some terms with which an expert would be familiar. An example occurs in the opening line of the text, where *planitie* (flat surface) is given instead of *palmo* (palm).

The text has been edited and translated by Charles Burnett, who has found no earlier version of the information in Latin and who concludes that this is not only the oldest known Latin text on chiromancy, but is also very close to the original source.[13]

It begins with the statement that there are three important lines to be found on every hand and that these form a triangular shape. Most significant of all is the Life Line, here also called the 'higher line', whose length corresponds (we are told) to the length of the individual's life. The other two lines are described in terms of their position in relation to the Life Line: one is called the Middle Line and the other the Oblique Line. It should be noted that this does

Figure 3.3. Detail of the chiromancy text. Cambridge, Trinity College, MS R 17 1, f282r.

not correspond to the identification of the three main lines in modern treatises on palmistry, and that later medieval texts show some confusion over the names, positions and relative importance of these lines. The Middle Line is very close to the Life Line at one end, and thus is presumably to be identified with the modern Head Line. Nothing is said at this stage about the prominent line which runs roughly parallel to this second line and closer to the bases of the fingers (now known as the Line of Heart). In fact, the lines in themselves are less significant than marks and shorter lines found on or near them. It is these latter 'signs' that are specified in detail and illustrated with small, graphic forms set in the text. The one most visible in Figure 3.3 is the sign like a large P, with a very small v resting on the upper stroke of its loop (P̌) which can be seen almost at the end of the ninth line of the left-hand column of text. We learn that the presence of this sign on a hand proves that its bearer is a virgin. More ominously, short lines cutting across the angle between the Life Line and Middle Line indicate various forms of death by violence or misadventure. Anyone with a mark like a letter c, but with a line cutting into it, placed near the end of their Life Line could rejoice in the news that they were destined for advancement and honours. If that sign was accompanied by another, looking something like a version of the number 11 in Hindu-Arabic numerals, then the news was very good indeed. These signs can be seen in the fifteenth and sixteenth lines of text above, shortly after the P-like sign.

As well as giving indications concerning future prosperity and the cause of death, the text also ascribes medical significance to some features of the lines. For instance, three short lines joined at one end and cut across by another line signifies a danger of leprosy.

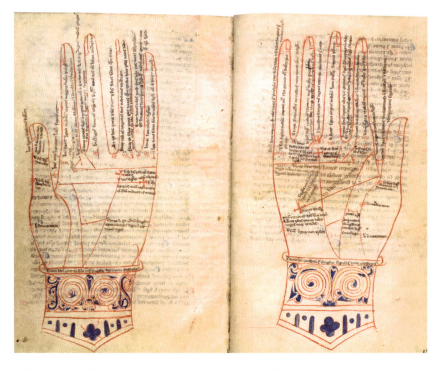

Figure 3.4. Illustration of chiromantic hands from a thirteenth-century manuscript. London, British Library, Sloane MS 2030, ff125v–126r.

Equally worrying is the appearance of red colour in the lower half of the 'first natural line', since that signifies serious pain in the head, while red in the other part of the line foretells pain in the neck. This may suggest that this 'first line' is the one later identified as the Head Line. Deep lines cutting across the Life Line and into the 'hill' below the thumb suggest strong feelings and a danger of uncontrollable anger. However, the presence of numerous lighter lines on the 'hill' is to be welcomed, since that indicates freedom from, or rapid recovery after, illness. Still other signs indicate children and their likely sex, while short lines at the bottom of the little finger or between it and the second finger suggest wounds and their bodily locations. The text ends with a list of individual signs and their significance, the last one being that a sign like a letter 'O' with a horizontal line crossing it indicates the loss of feet (shown in the right-hand column in Figure 3.3 above). Tantalisingly, this is followed by the statement that 'secret matters are dealt with in another place'. No indication is given as to what these secret matters might be, but it is tempting to wonder whether they might have been instructions on how to look for answers to questions about specific decisions, such as sending an army into battle. This abrupt ending does at least suggest that the 'signs' already covered could be considered natural and acceptable, rather than 'secret'. The fact that Augustine himself had acknowledged that doctors needed to read signs and make prognoses may well have been helpful, although it could hardly be applied to all the topics covered here.[14]

As is shown in Figure 3.3 above, the text on palmistry is followed by an empty space. The short instructions for using the technique known as 'onomancy', or signs and predictions to be derived from names and words, follow on the verso of the same folio. These are not written as continuous prose and cannot be used without pre-existing knowledge of the techniques involved. It seems likely that, as was common with textbooks on technical subjects in this period, both texts were intended as reminders to experts and teachers

– something which fits with the high status of this very special psalter. Charles Burnett has edited and studied the text on onomancy, as well as that on chiromancy, and his work will be used here.[15] The information provided makes it possible for the user to receive answers to questions on topics such as: the outcome of illness; which side will prevail in a contest of any sort; love affairs; and fertility.

The text starts with the days of the week (giving the name of the planet with which each day was associated) and assigns a numerical value to each day. Sunday (the Sun) has 24; Monday (Moon) 26; Tuesday (Mars) 5; Wednesday (Mercury) 20; Thursday (Jupiter) 19; Friday (Venus) 30; and Saturday (Saturn) 24. Next come nine 'distinctions' or sections, each setting out possible pairings of numbers from 1 to 9 (inclusive). Thus, the first section covers combinations involving the number 1, and how it will fare in relation to the appearance of 2, 3, 4, 5, 6, 7, 8 or 9 as the other number. For instance, if one party has 1 and the other 3, then the side signified by 1 will prevail. But if one side has 1 and the other 6, then the holder of 6 will prevail. Naturally, the sections grow shorter as they progress, since the lower numbers have already been dealt with. By the time 9 is reached, we simply learn that the combination of 9 and 9 signifies that the 'smaller' or 'weaker' side will prevail. The final part of the text shows which types of questions are expected, since it gives instructions on how to produce answers concerning: the outcome of illness; the outcome of combat or of any struggle between two opposing factors or sides; marriage; and love (the last by extension). At the bottom is a list of the letters of the alphabet and the number values assigned to each – crucial if names (or other verbal information) are to be turned into numbers for the requisite calculation. A and B are both valued at 3; C is 22; D 24; E 22; F 3; G 7; H 6; I/J and K 15; L 12; M 23; N 15; O 8; P 13; Q 21; R 13; S 9; T 8; U/V/W 6; X 5; Y 3; and Z 4.

Taken together, this means that answers can be given to questions such as whether and when a marriage between two people should take place; which party in an argument is likely to prevail; or whether a sick person will recover. Later versions of the Latin text provide further information by giving instructions for using the numbers in specific cases, but the 'Eadwine' text offers only the essentials. The most straightforward application of onomancy is the 'combat' between two sides. For instance, where two individuals are in conflict, the letters of their names can be used to produce two numbers. If one is William and the other Robert, then the sums produced (using the Latin versions of the names) would be respectively 105 and 82. Each number is then to be divided by 9, as directed in the final section of the text, and the remainder is to be noted. In this case, the remainders are 6 and 1. As shown by the information given above, this means that the side represented by 6 (William) will triumph. Clearly, the same procedure can be followed wherever 'names' can be applied to two competing factors or sides. The calculation for a sick person is simpler than might be expected, since the numbers used are based on the patient's name and the day on which they fell ill; thus, there is no requirement to give a name for the illness itself. Both of the numbers produced are again to be divided by 9 and the two remainders again to be assessed using the lists of pairings. A woman named Matilda would have the number 88 and the remainder 7. If she first became ill on a Friday, the number would be 30 and the remainder 3. In such a case it is the side represented by 3 (i.e. the illness) which is stronger, which would not be good news for Matilda. In questions involving married couples, the procedure is the same, but the divisor is 7. It should be noted that the text assumes that of the two final numbers one will always prevail, even in cases of marriage (or, possibly, love).

The procedure of turning names and dates into numbers, and of using a text claiming ancient origins to deduce an outcome, was already established in medical practice in medieval England and Europe from the ninth century onwards. The best-known example was the short text and diagram known as 'Sphere of Apuleius' or

'Sphere of Pythagoras', which appeared in western European manuscripts from the ninth century onwards. This not only promised insight into the outcome of illness, but also made vague reference to possible consultation on other matters; however, the accompanying instructions focused on medical matters. The procedure was to take the letters of the patient's name and to find their numerical value in the table provided. The day of the week on which the patient fell ill should also be given a numerical value, and finally the day of the lunar month should be noted. All these numbers should be added together and the total divided by 30. The remainder should then be found within the circular diagram (the 'sphere' of the title). If it was located in the upper part of the circle, the news was good, although the exact location could modify this. Equally, a placement in the lower half of the circle was an indication of danger of death.

This early reception of the 'Sphere of Apuleius' is important both because it meant that the technique involved in the newer and more general form of divination was already familiar and because the manuscripts in which it is found were produced and used in monasteries. These manuscripts frequently included texts and tables that were part of the monastic science of *computus*, which focused on the calculation of time and the significance of its passage and components. The allocation of meaning and effect to times and dates was central to this science, as was the performance of complex calculations related to the dating of 'moveable feasts', such as Easter. These involved such factors as the date in the lunar calendar, the position of the Sun in the zodiac and the place of any given year in the nineteen-year cycle used to correlate the lunar and solar calendars. The concept of allocating numerical values to letters of the alphabet was not central to *computus* itself, but nevertheless lists of alphabetical symbols and numerical equivalents are found in some *computus* manuscripts. A well-known example is the twelfth-century volume from Thorney Abbey, now Oxford, St John's College, MS 17.[16] Folio 5v of the Thorney manuscript provides an impressive set

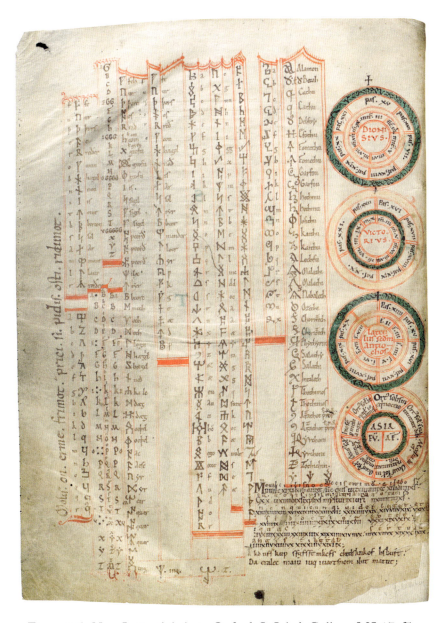

Figure 3.5. Non-Latin alphabets. Oxford, St John's College, MS 17, f5v.

of non-Latin alphabets, set out in parallel and with a handy table in the bottom right-hand corner.

Older versions of the explicitly divinatory text found in Eadwine's Psalter exist, but they are in Greek or Arabic. Burnett's research found lists of numerical values for letters in the collection of short astronomical and astrological texts known as the *Liber Alchandrei* which circulated in northern Europe in the eleventh century. Like other books offering new and strange knowledge, the *Liber Alchandrei* linked much of this information to Alexander the Great and his tutor, Aristotle. However, these lists are different from the ones given in the Eadwine Psalter and are clearly not its direct source.[17] The assertion that Aristotle was responsible for drawing up lists of numerical values for letters, and tables for using them, is more direct in Arabic works, where such information is often incorporated into volumes offering (unsolicited) advice to rulers. Aristotle was believed to have been the tutor of Alexander the Great, and to have imparted secret knowledge which helped Alexander's extraordinary achievements. Supposed extracts from letters written by Aristotle to Alexander were well received in medieval Europe and were incorporated into a composite work known as the *Secret of Secrets*. Demand was sufficient for them to be translated into Middle English and other vernacular languages in the later middle ages.[18] However, as Burnett shows, the 'Eadwine' text could not be derived from any of the examples which survive. Thus, like the chiromancy text, this is evidence of very unusual expertise being present and valued in the monastic community at Canterbury Cathedral. The fact that it was placed on record in this particular volume, alongside the portrait of Eadwine and the impressively detailed diagram of the new water system, makes it important to look a little further at the manuscript and its place in the cathedral priory.

The fact that major work on the supply and circulation of water to the cathedral precincts was undertaken at the time the manuscript was being produced has already been mentioned. This followed important building works on the cathedral itself. The much-admired new choir, constructed in a fashionable style, was dedicated with magnificent ceremony in 1130. Prior Conrad (1108–26) had searched Europe and the Mediterranean for artists and art works to decorate this new choir. Subsequently, Archbishop Theobald (1139–61) recruited ambitious clerics and scholars to work in his administration.[19] This would bring together an impressive group of talented individuals within the precincts of the cathedral. Such activities were expensive, but Prior Wibert (1152–67) is credited with overseeing a revival in both the fortunes and the splendour of the priory.[20] The attention of historians has focused on the magnificent buildings created at Canterbury during this time, and on individual art works such as Eadwine's Psalter itself; but there is evidence that the creation of the latter formed part of a campaign to overhaul and update the most important manuscripts of the cathedral.

The cathedral and its priory owned a collection of beautiful and historic manuscripts, which already included at least two impressive psalters. The one most closely linked to the Eadwine Psalter is now Harley MS 603 in the British Library (and consequently usually known as the Harley Psalter). Unlike Eadwine's impressive volume, the Harley Psalter predominantly follows just one Latin version of the psalms – the one referred to as the Roman, which was used in pre-Conquest England. This is surprising, since the striking illuminations of this manuscript, and its page layout, are closely based upon a Carolingian psalter which uses St Jerome's first translation of the psalms, known as the Gallican version. The Carolingian manuscript is now in the library of Utrecht University and is consequently known as the Utrecht Psalter. It seems to have arrived in Canterbury at the end of the tenth century and to have remained there for several centuries. Both the Harley Psalter and the Eadwine Psalter are updated versions of the Utrecht Psalter. Harley 603 was begun early in the eleventh century and contains script and

illuminations by several artists and scribes who produced manuscripts for Christ Church and its priory. For reasons that remain
unclear, the Harley Psalter was left incomplete. However, at the
same time as work was under way on the Eadwine Psalter, a short-
lived attempt was made to fill some of the gaps in the illumination
of the Harley Psalter. These focus on a fairly early section of the
manuscript, folios 29 to 35; the missing later part of the text seems
never to have been completed.[21]

The sibling relationship between the Harley and Eadwine
Psalters might suggest that the attempt to fill gaps in Harley was
simply a by-product of the campaign led by Eadwine. However, the
evidence provided by the volume known as the Arundel Psalter
contradicts this. This manuscript is now Arundel MS 155 in the
British Library and is mostly the work of Eadui Basan, a prolific
and celebrated scribe-artist who produced manuscripts for Christ
Church in the early eleventh century – and who made small contributions to the Harley Psalter as well.[22] In the Arundel Psalter, the
main psalm texts were copied by Eadui and extend up to folio 135r.
Folios 171 to 193 also show Eadui's distinctive style, and they
contain prayers, texts for use in confession and Old English glosses.
However, folios 135v to 170v are very different in appearance and
contain work by several mid-twelfth-century scribes. Their texts
include a litany, canticles, prayers, hymns and the Office of the
Dead, which would bring the volume up to date for liturgical use on
solemn occasions. The effect of these additions is thus to make the
Arundel Psalter something rather different from either Harley or
Eadwine, which were unsuitable for such use. This suggests a ratio-
nalisation of this impressive set of psalters. Arundel would be cere-
monially used on major religious feasts, while the Eadwine Psalter
displayed the cutting-edge knowledge assembled at Christ Church,
as well as providing a fully updated version of the Utrecht Psalter.
The Harley Psalter could thus remain incomplete, while still apparently being valued and kept.

If that is the case, then this analysis suggests that Eadwine's Psalter was mainly intended for academic study. This accords with its intense focus on the words of the psalms, their equivalents in several languages and the rather literal, visual interpretations of key phrases from psalm texts. The latter were very unusual, and were directly based on the Utrecht Psalter. Perhaps equally striking are the details of classical buildings, sculptures and clothing included in the images, which transport the psalmist and other figures into an imagined version of the ancient world. This is emphasised still more by the use of Roman-style 'rustic' capital letters in the Carolingian manuscript. The illustration below (Figure 3.6) shows the dream-like images which result.

This original can usefully be compared with the versions in the Harley Psalter and the Eadwine Psalter. The handling of text and image in the Anglo-Saxon manuscript is clearly close to the Carolingian original, and yet the use of colour and of the 'modern' script known as Caroline minuscule makes this recognisably something different.

In the case of the Eadwine Psalter, the opening of Psalm 58 has been updated again in terms of its visual appearance, while still being recognisably based upon the design and layout of the Utrecht original. This version is less obviously uniform in its layout and less dominated by the illustration; but the result is that great emphasis is placed on the psalm text and its interpretation. In other words, this manuscript combines ongoing interest in the unique illustrations found in Utrecht with a new and impressive body of scholarly work on the text. To take just one example, analysis of the Old English glosses and their corrections has produced very interesting results. The glosses accompany the Roman version of the psalms which, as already noted, was the one most used in pre-Conquest England. Linguistic analysis has found that these glosses combine archaic and idiosyncratic Old English words with more modern ones.[23] The archaic glosses have not been found in any other

Figure 3.6. Opening of Psalm 58. Utrecht, Utrecht University Library, MS 32 (Utrecht Psalter), f33r.

surviving manuscript and contain errors, suggesting that they were originally entered into a manuscript that was hard to read. This has led to the suggestion that the glossed manuscript used as an exemplar was even older than the Harley Psalter and was presumably prized for its antiquity.

Figure 3.7. Psalm 58. London, British Library, Harley MS 603
(Harley Psalter), f32r (on the right).

That leaves the inclusion of the divinatory texts still rather
puzzling, but not as much as they would have been in a less academic
volume. The idea of Eadwine's Psalter as a focus for 'modern' scholarly work also places the project in still another context within
twelfth-century Canterbury – that of Archbishop Theobald's international group of clerics, scholars and scientists.[24] It seems likely
that at least one of them was an expert on Hebrew. The evidence for
this is that the Eadwine Psalter, uniquely among English manuscripts from this period, has both accurate versions of Hebrew
letters and evidence of understanding them. Its copy of Psalm 118,
found on folios 211r to 227r, has the letters themselves in the headings for each textual section, placed close to the correct versions of
their names. These could not have been copied from either the
Utrecht or the Harley Psalters.[25] It should also be noted that such
expertise was received with suspicion in some quarters in the earlier
part of the twelfth century. Archbishop Gerard of York (d. 1108)
owned two Hebrew psalters, and this contributed to rumours that

Figure 3.8. Psalm 58. Cambridge, Trinity College, MS R 17 1, f101v.

he was interested in the dark arts, although his manuscripts were retained at York and were consulted by the scholar Maurice of Kirkham towards the middle of the century. Meanwhile, in Paris, the School of St Victor included a group of scholars studying Hebrew, at the same time as Herbert of Bosham, the friend and supporter of Thomas Becket, was studying theology under Peter Lombard. Herbert of Bosham also spent part of his career in Canterbury, where, as well as writing on theology, he penned a commentary on St Jerome's translation of the psalms from a version in Hebrew.[26]

This range of interests and collection of experts helps to show just how much scarce knowledge was brought together and drawn upon in mid-century Canterbury. There is no direct mention of 'suspicious' individuals in contemporary sources; but John of Salisbury's critical comments on the fashionable status of the dark arts in England, and their patronage by Becket in particular, start to look much more understandable. The idea of access to long-lost and scarce knowledge from the ancient world was very attractive in the twelfth century and was becoming more possible to attain – at least for the elite. A final point linking the text of the psalter to instructions for finding signs of the future may perhaps be seen in the unusual preface to the psalms found in the Eadwine Psalter. This opens with a resounding statement on prophecy, its origins in divine revelation through various routes and comments on examples of true prophets. The ultimate source of the statement is an authoritative, sixth-century commentary on the psalms. However, this endorsement of prophecy and true knowledge of the future forms a striking introduction to a volume that concludes with texts offering such knowledge through (in the words of the preface) 'many and various means'.[27]

This analysis has suggested that the 'new technologies' for gaining insight into the future that were proudly copied into Eadwine's Psalter occupied a somewhat equivocal position in the manuscript overall. On the one hand, they are integral to the concluding section of the manuscript and to the achievements and innovations recorded there. On the other, they are additions to the main text of the psalms, and their relationship to Eadwine himself is left unexplained. Similarly, while Eadwine's skills as a scribe and creator of this impressive book are celebrated in the highest possible terms, no information is given about the man himself. It is tempting to suggest that he might have been a recent convert to the monastic life; but that can only be speculation. The next manuscript to be considered offers a complete contrast on both those counts.

The book, which is now Ashmole 304 in the Bodleian Library, Oxford, is a divinatory device in itself and consists entirely of a set of complex and beautifully illuminated *Sortes* tables, used for the procedure known as 'casting lots'. These were originally accompanied by at least one rotating dial (volvelle).[28] The manuscript has been universally identified by scholars as the work of Matthew Paris, monk of St Albans and best known as the creator and illustrator of impressive and influential chronicles and saints' Lives. Matthew joined the St Albans community at an early age; but, once he became a fully professed monk, he appears to have been given permission to spend almost all his time on his literary and artistic productions.[29] Moreover, Matthew himself proudly records the occasions on which he received the honour of personal conversations with the most powerful people in the kingdom. A few of these meetings took place in St Albans itself, but Matthew also travelled to join the royal court at Westminster and other centres of power – something extremely rare for a monk.

Matthew never took on an official role within the hierarchy of the abbey, despite his clear ability. The works for which he is now best known are his chronicles, which draw upon much better sources of contemporary information than was usual. This information seems to have been derived from a network of contacts, which brought this cloistered monk to international attention in 1247.

This was the year in which King Haakon of Norway asked Matthew to mediate with moneylenders from Cahors, in France, at the request of the monks of the Norwegian abbey of St Benet Holm, Nidarholm. It seems that, for reasons which he does not divulge, Matthew was already familiar with the business methods of these international financiers, and that he was perhaps personally acquainted with one of the group based in London.[30] In this same year of 1247, Matthew also recorded that he attended the ceremony at Westminster Abbey when an impressive relic of the Holy Blood was received in England, sent to King Henry III by the patriarch of Jerusalem.[31] He seems to have been invited by the monks of Westminster, who already knew of his position as chronicler at St Albans; but on this occasion he was singled out also by the king himself. If Matthew's account can be believed, Henry III made him almost the historian of the court, since he instructed him to record what he observed, and even invited him to dine at the royal table.[32] It is clear that the contact continued, since Matthew records further conversations and exchanges of gifts and books not only with the king, but also with the queen, members of the high nobility of England and royal officials.

Those in high places who talked with Matthew appear to have been impressed by him. This emerges clearly in the case of King Haakon, Nidarholm Abbey and the Cahorsin financiers, mentioned above. The story is complicated, but began when the abbot of Nidarholm sold off most of the abbey's assets, before absconding to the papal court in Rome and taking the abbey's seal with him. This he used to borrow money from the Cahorsins, thus worsening the financial position of the monks, who accused him of serious misconduct. It was because they knew that he had previously made a favourable impression on their king that the Norwegian monks chose Matthew Paris to negotiate on their behalf: Matthew had earlier delivered to Haakon some letters from King Louis IX of France, inviting Haakon to take part in the crusade that Louis was working to organise. Haakon had declined the invitation, but gave Matthew

Figure 3.9. A possible self-portrait and personal
prayer by 'Brother Matthew of Paris' in a manuscript
containing the last part of his *Chronica Maiora* with
his *Historia Anglorum*. London, British Library, Royal
MS 14 C VII, f6r.

privileged information – just as other European leaders had done,
including Richard of Cornwall, younger brother of Henry III, and
Eleanor of Provence, the queen of England. Matthew was privy to
the fact, for instance, that Henry III maintained and consulted an
astrologer at his court. And he was even able to record that when
Henry was anxiously watching events in France in 1226, his decision
not to invade was reinforced by advice given by William of Pierrepont,

his highly skilled expert in the science of the stars. According to Matthew, William had reassured Henry III that Louis IX's crusade would not go well for Louis himself.[33] It might be expected that a monastic chronicler would be critical of such use of astrology; but in fact Matthew describes William's skills in a very positive way,

Figure 3.10. The terror of the conquest of England by the pagan Swejn. Queen Emma flees on horseback, with her young sons Alfred and 'Eadwardulus' or 'Little Edward'. Cambridge, Cambridge University Library, MS Ee 3 59, f4r.

treating astrology as a serious branch of knowledge – one that required special training. There is also a suggestion that Henry not only took William's predictions seriously, but was right to do so.

Of course, members of the court will have wished to appear in Matthew's chronicle in a favourable light – but that assumes they were already aware that Matthew's work would achieve a special position. It has indeed been accepted as the most important of those chronicles being written in the religious houses of England, precisely because of the 'inside information' given to Matthew. From about 1250 onwards, Matthew's account of English and European history became both more privileged and less critical of the king. At the same time, copies spread across southern England, from London to Winchester, Norwich, Canterbury and Rochester, as well as to Tintern in Monmouthshire (to list only those that survive). Thus, Matthew's achievement as an 'official' historian is undeniable – and this raises the question of how he achieved this highly unusual status. His skill in negotiating with bankers was clearly useful, though it was hardly enough on its own.

It may have helped that Matthew was able to provide beautifully produced and illustrated volumes of the Lives and miracles of saints for the great ladies with whom he was in contact. His Life of Edward the Confessor was composed for Queen Eleanor, to whom it is dedicated, and was written in Anglo-Norman. Matthew's original presentation copy does not survive, but the volume now MS Ee 3 59 in Cambridge University Library is accepted as a close copy.[34] Its illuminations, like those in other manuscripts produced by or linked to Matthew, are impressively composed and carried out in a fashionable style and technique. A little later, Matthew also wrote a Life of St Edmund, which was presented to Isabel, countess of Arundel, as was a book of verse with pictures. The latter was presented to a countess of Winchester, probably Maud, the widow of Anselm Marshal (to whom Isabel of Arundel was also related). Matthew clearly retained illustrated copies of his works, since he recorded that his Lives of both Edward and Thomas were lent to both Isabel of Arundel and the countess of Cornwall (wife of Richard, King Henry's brother).

Also highly unusual, and of considerable interest to those in high places, are Matthew Paris' detailed maps of Britain and the route to the Holy Land. These demonstrate both up-to-date techniques and information that must have been carefully compiled from a number of sources.[35] Close links to the royal courts of both Henry III and his son, Edward I, have been identified in these maps.[36] It has been convincingly argued that they bring together geography, history and prophecy. Matthew's interest in prophecy was far from unique – but he incorporated into the St Albans chronicle a selection of recent prophecies, placed in close relationship to his accounts of major events. Specifically, he cited both the prophecies of the Sibyl and those of the Cistercian prophet, Joachim of Fiore – although he noted the criticisms of the latter.[37] Joachim had applied textual critique to books of the Bible, particularly the Apocalypse, and based both guidance for his own time and prophecies of the future upon this. He died in 1202, but was widely believed to have foreseen the imminent coming of Antichrist, bringer of the End Times; and some commentators supported the precise date of 1260 for this event.[38] Matthew died in 1259 and so did not live to see whether the prophecies were fulfilled; but both the illustrations and the comments in his chronicles demonstrate the strength of his interest in Joachim's revelations. For a chronicler with Matthew's range of knowledge, the belief that the patterns of human history could be revealed by an inspired reading of the Bible was extremely interesting – not least to his powerful patrons.

This discussion has suggested that Matthew's production of a multiple divinatory device, in the form of a beguiling book, fits better with his other works than has been recognised. Not only were Matthew's patrons interested in prophecy and astrology, but (as the first part of this chapter showed) they and their predecessors also expected to be able to consult experts within the Church on questions relating to the future. What remains striking is the combination of beauty and mystery that the book displays. It is no longer in its original binding and has lost its opening folios, so at first sight it is not as impressive as it must have been when first made. However, it was always designed to be portable, even in a luxurious binding and with space for its revolving dials. It now has only seventy vellum leaves, and the experience of opening the manuscript is less impressive than it would have been originally. Nevertheless, the first main text still commences with a scene-setting miniature in Matthew Paris' fashionable style, depicting two famous and respected experts in the science of the stars – identified as Euclid and Herman. Euclid, representing the knowledge of the ancient world, is shown observing the Moon through a viewing tube (not a telescope), while holding up a small armillary sphere (a model of the universe, with the Earth at the centre and rings depicting major elements, such as the belt of the zodiac). On the right of the image is the 'modern' figure of Herman, holding out the new technology of an astrolabe, which is prominently labelled in red. Knowledge of this Islamicate instrument had reached northern Europe in the early eleventh century, and a monk known as Hermannus Contractus or Herman of Reichenau (c.1013–54) played an important role in transmitting texts and instructions.

This image, with other parts of the introductory materials in the volume, serves to link it to cutting-edge discoveries about the celestial bodies and their movements. The stars and planets were accepted as exerting strong influences over all material things on Earth, including human bodies, atmospheric phenomena, winds and ocean tides. It was therefore logical to argue that accurate knowledge of their movements and positions, linked to expert use of the astrolabe, could provide foreknowledge of coming influences and effects. This was the basis of the respect accorded to highly skilled astrologers like William of Pierrepont, and it is important to note that Matthew Paris does not claim this type of expertise for himself. Instead, the

Figure 3.11 Euclid and Herman. Oxford, Bodleian Library,
Ashmole MS 304, f2v (detail).

divinatory tables and symbolic images offered in his book provide a
long-established form of shortcut to knowledge of the future –
which is here updated by new links to astrology and the insertion of
references to the Islamicate experts whose works had been translated into Latin in the twelfth century. It is the combination of a

wide-ranging set of the divinatory tables known as *Sortes* or 'Lots', the references to classical and Islamicate astrology and geomancy, the beautiful and elaborate images, and the new techniques for consulting the *Sortes* tables that makes this manuscript both fascinating and dizzying.

A semblance of the experience of a consultation with the book – via the Master who understood its use and delivered its responses to the Client – can be gained even now, as a fourteenth-century copy of Matthew's manuscript is now Digby MS 46 in the Bodleian. The book opens to reveal something that was very new in the middle

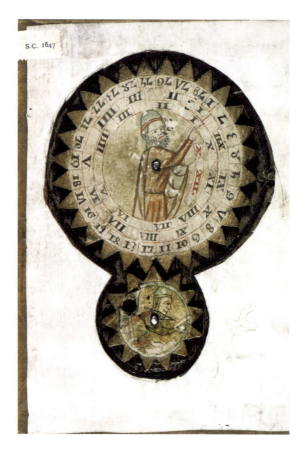

Figure 3.12. Double volvelle inside upper leaf of binding.
Oxford, Bodleian Library, Digby MS 46.

of the thirteenth century – a volvelle. In fact, Matthew Paris' volvelles, which are found also in the maps and calendar materials attached to copies of his Great Chronicle, are the earliest from western Europe known to survive – thus providing further evidence of the research and expertise that this erudite monk accumulated.[39] In this case, the smaller, lower wheel contains the figure of a man bending to point at the teeth around 'his' wheel, while holding a written page in his left hand. This wheel also has a small hole, into which a peg or stylus could be inserted. The main function of this small wheel is to make the larger, upper wheel rotate in the opposite direction. The upper wheel contains a profile figure of a well-dressed, grey-bearded Master, holding in his left hand a scroll on which are marked the Roman numerals X and XII. His right hand points to the outer part of the wheel, where three concentric rings of numbers can be found. Two of these are in Roman numerals, and run from I to X and from I to XII, respectively. The Master's right index finger is placed at the point where these cycles start and finish. Beyond them, and starting at the same place, is a ring of Hindu-Arabic numerals, running from 1 to 28. The Master depicted here is not named, and his volvelle could be used in the consultation of several tables within the book; but there is also evidence that other volvelles originally existed, linked by image and name to specific texts and tables. For instance, the *Prognostications* (or prophecies) of Pythagoras were originally preceded by a device bearing his image. This is mentioned in the instructions for its use, which are included on folio 40v. A larger image of Pythagoras, shown writing the text itself, is on folio 42r, although the intervening leaf and volvelle have been removed.[40]

The sets of numbers in the double volvelle correspond to the sets of questions offered in different tables within the book, with the most 'modern' ones linking to the fullest, and most modern, tables – as will be seen. It is first worth pointing out that, if Hindu-Arabic numerals were also used in Matthew Paris' lost original of this device, then it was even more impressive in its display of new technology than first appears. Here divinatory magic is being presented both as an inheritance from the classical world of Euclid and Pythagoras and as a means of harnessing the new astronomical and astrological knowledge freshly translated from Arabic – and in many ways that was a fair claim.[41]

The consultation would begin with the selection of an appropriate oracle from the range offered by the book. Those looking for guidance on matters touching on moral or ethical dilemmas might perhaps select the Lots of the Twelve Patriarchs, while those seeking more secular advice could appeal to the wisdom of Socrates, Pythagoras or Cicero. The loss of the opening folios of the first main text has made its 'oracle' rather ambiguous, but the twelfth-century philosopher and poet Bernardus Silvestris is named as the translator. That is slightly odd, since it is also claimed to be the work of a doctor working for 'King Amalric'. This would be Amaury I, king of Jerusalem (r. 1163–74), whose doctors might be expected to have produced texts in Latin.[42] However, the doctors of such a king would also have been expert in the Islamicate sciences of astronomy, astrology and medicine, invoked when the preface to the text in Matthew's version cites the famous name of Albumasar (Abū Ma'Shar). In fact, analysis of the text by Charles Burnett has shown that the author of the instructions for consulting the work, and the compiler of the tables attributed to the anonymous doctor, did not understand Arabic.[43] The resulting confusion in the text was serious enough to be embarrassing if users of the book understood Arabic and were familiar with the Arabic names for the lunar mansions – but the risk of that happening in thirteenth-century England and among users of Matthew Paris' book was very low.

Assuming that clients seeking answers to a troubling question chose to look for guidance to the planets, and to the Moon in particular, the Master guiding them through the book would presumably explain the claims made in the preface that this proce-

Figure 3.13. Table of questions. Oxford, Bodleian Library, Ashmole MS 304, f3r.

dure would produce enlightenment as to the coming effects of the planets and the stars. The claim that this is an astronomical treatise, translated and guaranteed by Bernardus Silvestris, would presumably be included – and supported by pointing to the diagrams of star-groupings shown at the bottom of folio 2r and labelled 'Constellationes'. The fact that these diagrams are not only 'constellations', but also illustrations of the equally fashionable groupings of dots produced and interpreted during geomantic consultations would add a further element of predictive glamour.[44] The layers of authority deepen when the leaf is turned to reveal the miniature of Euclid and Herman already discussed. The Masters are seated on richly carved benches, and below their feet are the instructions for the first crucial stage in the path from question to answer. This is that, as in a geomantic consultation, the Enquirer should make a row of dots, taking care not to count them during the process.

Once the pen has been put down, the dots should be counted and note taken of whether the number is higher or lower than seven. The Enquirer then goes to Table 1, consisting of twenty-eight questions fitted into columns of coloured text on folio 3r, and chooses the one most appropriate. For instance, those contemplating a long or difficult journey might choose 22 'De Via' (the Road). The line of that question is then counted as having the number 7 for the purposes of this consultation. If the Enquirer made, for instance, a row of twelve dots, they should then move up a further five lines to reach their total of twelve (a number lower than seven would lead them to move down the lines to the one indicated). Moving up five lines will take them to the line corresponding to Question 17, and by following that line across the columns of text the Enquirer will learn that they should seek the eastern face of the Tower of Mercury. This is explained by turning the leaf, as Table 2, which sets out the Towers of the Seven Planets, placed on folio 3v. Mercury's tower is positioned third from the bottom, with its 'eastern face' occupying its first line of text. This in turn directs the Enquirer to seek the twenty-fifth lunar mansion.

No explanation of this idea is given, although careful instructions and labels have been supplied for much of the material. In fact, as Charles Burnett has pointed out, the reference is to the twenty-eight lunar mansions (or Mansions of the Moon) adopted by Islamicate astronomers.[45] These are divisions of the ecliptic (the apparent path of the Sun around the Earth in a geocentric universe), each equating to the distance covered by the Moon in a period of twenty-four hours. They thus add up to a complete circuit of the zodiac by the Moon, taking almost twenty-eight days, and were usually defined both by the stars and star-groupings linked to them and by their relationship to the divisions of the zodiac allocated to the astrological signs. The first mansion was usually taken to be placed in the first portion of the zodiac sign of Aries. The knowledge of these lunar mansions, with transliterated lists of their names, reached Latin Europe as part of the short astrological and astronomical texts associated with the name Alchandreus. It is clear from the various terms and descriptions applied to these lists that they were a new and unfamiliar concept, and it is not surprising to find that their names occur in Latin texts in varying forms. They are treated in Ashmole 304 both as concepts linked to the Moon and as Judges delivering answers to questions asked via the tables in this 'astronomical' text.

The mystery deepens as two further stages of redirection follow, at the end of which the Enquirer is instructed that the answer will be given by the twenty-third 'Judge' named *Cateateve*. The judgements, under the name *Catheacene*, are to be found on folio 28r. It is worth noting here that these are indeed versions of the name of one of the lunar mansions – although it is in fact the twenty-second mansion, not the twenty-third.[46] Still following the introductory Instructions, the Enquirer now goes back to the original number of dots (12) and this time applies the fixed number 9. Adding 9 and 12

Figure 3.14. The Towers of the Seven Planets (detail). Oxford, Bodleian Library, Ashmole MS 304, f3v.

produces the number 21, and the Enquirer duly consults the twenty-first line of Latin verse on folio 28r. This reads *Cum da[m]pno vadis nam plena sit h[uius] via cladis*. In other words, the projected journey, if undertaken, will be destructive, being full of trouble and disaster.

Two things follow from this. The first is that the Enquirer might well be grateful for having been given such a strong warning and for the opportunity to change their plans. The second is that, despite all the tables and redirections intervening between question and answer, the judgement does indeed directly relate to a journey. The muddle over the names and correct sequence of the lunar mansions makes no difference to the working of the algorithm linking questions to tables and finally to answers. Like several of the 'Lots' texts themselves, this algorithm goes back to the late antique period, although increasing the number of intervening tables also increases the complexity of the calculations that the designer of tables is required to make. Put simply, there are usually as many questions as there are Judges, and each question will need to have at least as many possible answers as there are possible Judges to deliver them. It should also be possible for each Judge to answer each question. Given those requirements, the placing of the answers within the table allocated to any Judge should not be obvious – there would be little need to bother with intervening stages if the answer to Question 22 was numbered 22 in every case. All that would be required is the choice of Judge. However, the placing clearly cannot simply be random, unless all the answers are to be couched in disappointingly general terms. The complex procedure for shuffling the sets of answers, and for making sure that questions will lead to relevant answers, is discussed in detail by Michael Meerson.[47] It also requires the placing of dummy answers in the Judges' tables, to compensate for the addition of nine at the final stage.

This is the largest and most complex set of tables in Matthew Paris' manuscript, as well as the one that makes most reference to recently translated texts of astronomy and magic; and it is clearly no coincidence that it has precedence within the book. However, the sheer range of divinatory tables collected here is impressive testimony to the work and research that went into the volume. While many of the tables can be traced back to late antique precedents, it is still the case that Matthew Paris' versions stand early in their traceable transmission into Latin Europe. Moreover, Matthew's tables of Lots are more complex – in terms of both the procedure for consultation and their ability to answer very specific questions – than the versions available in Latin Europe prior to 1200.

The dominant earlier form of divination by Lots was the procedure now known as the *Lots of the Saints* (*Sortes Sanctorum*) or *Lots of the Apostles*. Known examples, including in Anglo-Saxon and Anglo-Norman manuscripts, are most often associated with texts on astronomy and the calculation of the Church's calendar. One such, which has already been mentioned, is the 'encyclopaedia' of computistical texts produced in Thorney Abbey in the twelfth century and now MS 17 in St John's College, Oxford. The text is found on folios 157v–158r and is given no title, although a space is left on the first folio. What is presented here consists of a list of short predictions, written in black ink, each following a group of three numbers written in a non-standard form of Roman numerals and in red ink. The list opens abruptly with 6-6-6, and proceeds down to 1-1-1. There are no instructions, but it is fairly clear that dice-rolls are the most likely means for the Enquirer to produce the

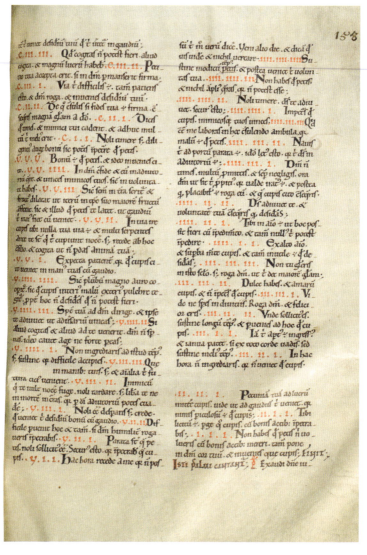

Figure 3.15. Part of the *Lots of the Saints*. Oxford, St John's College, MS 17, f158r.

three numbers. The answer always immediately follows the numbers, with no intermediate steps and no 'Judges'. Care is taken to invoke divine help and judgement, as in the response to 5-5-4, which states: 'trust in God and ask for His help and you will defeat your enemies as you wish'.[48] This provides helpful evidence for the view that the provision of such guidance was an accepted part of the functions of monastic experts; but such simple texts cannot have provided exemplars for those offered in Matthew Paris' volume.

A possible source for the St Albans versions of 'Lots' is suggested by their connections with both the ancient Mediterranean and the Crusader states. This is supported by the impressive range of information that went into Matthew's chronicles and maps. A case in point is the text entitled 'Lots of the Twelve Patriarchs'. This is placed fourth in Ashmole 304, on folios 52v to 55v, following the 'Predictions' of 'King Socrates' and those of Pythagoras. It is a simpler text, with twelve questions matching its Twelve Patriarchs – who are claimed to be the Judges delivering the answers.[49] Older copies of the questions and answers are known, including a version probably dating to the late eleventh century and from Salisbury Cathedral. This Salisbury manuscript contains predominantly texts on the calculation of the church calendar and on the tracking of the solar and lunar cycles in relation to religious feasts.[50] It does not link the Twelve Patriarchs to the lists of questions and answers, and in fact opens with a pious warning about the consultation of *Sortes* and a recommendation to prepare very carefully before undertaking such a thing.

Matthew Paris' handling of the text is very different. The Twelve Patriarchs are illustrated on folio 52v, and each is given distinctive features, as well as an appropriate name. They are also identified as the sons of Jacob and as the Judges who will give true responses. As the image below shows, the twelve questions are then set out, in compressed form, in the table below the miniature; and each question is immediately followed by the name of the relevant Patriarch.

A further biblical note is added when the responses are called 'Books of Prophets'. For instance, the question as to whether a journey will prosper is here Number 3, and the Enquirer is immediately directed to the 'Book of the Prophet Gaad'.

Gaad is the third Patriarch depicted in the miniature, and his answers are accordingly given third place in the sets of verses which follow. They are located on folio 53v, and Gaad is here specified as the Patriarch who gives answers concerning journeys. No

Figure 3.16. The Twelve Patriarchs. Oxford, Bodleian Library, Ashmole MS 304, f52v.

preliminary instructions are given, and there is no sign that a volvelle has been removed from the manuscript at this point. It thus seems likely that the Enquirer will have used the double volvelle at the start of the book, consulting the circle which gave numbers from I to XII. Exactly how that was done is not clear, but one possibility could be that the relevant number was the one occupying the position to which the Master's finger originally pointed. For instance, if that number is taken as also 3, then Gaad's response is a disappointing warning that efforts will be in vain.

Matthew Paris' decision to identify the Twelve Patriarchs as the Judges in these tables may have been an innovation, and it appears to reflect his interest in the separate text known as the *Testaments of the Twelve Patriarchs*. He records in his Chronicle that Robert Grosseteste, bishop of Lincoln, had sent John of Basingstoke to Greece to acquire this text, among other things. The Greek version was then translated into Latin by Grosseteste with the help of Nicholas, a Greek priest who, Matthew says, was part of the community at St Albans. It may have been through the intervention of Nicholas that Matthew was able to borrow the translation and make a copy, celebrating the prophetic powers of the Patriarchs in his Chronicle. The *Testaments* and the *Sortes* are quite different texts, although linked by Matthew through the emphasis on the patriarchs as prophets.[51] It is also possible that Robert Grosseteste provided Matthew Paris with information about (and a version of) the 'Lots of King Amaury'. The evidence for this survives in Oxford's Bodleian Library as MS Savile 21, which contains various computistical and astronomical texts, some in the handwriting of Grosseteste himself. In this version, the tables are preceded by a long introduction discussing the powers of the planets and giving details of the lunar mansions, before setting out a very long-drawn-out procedure for the actual consultation. This is perhaps more theoretical than practical, since it requires the different stages to take place on the appropriate days of the week and of a lunar month, which could spread over more than thirty days. What emerges is that the procedures involved in using the Lots seem to have been accepted. Grosseteste's interest in them is demonstrated by the fact that he copied extracts and notes relating to astrology and prediction on the verso of the last folio of the work.[52]

The final point to make about Matthew Paris' collection of divinatory tables is that it could cater also to those who believed that phenomena from the natural world, such as birds, could provide warnings and insights concerning the future. This, too, had been a topic for discussion in twelfth- and thirteenth-century England, with a well-known case being that of St Hugh, an earlier bishop of Lincoln, and his 'marvellous' swan. Both Gerald of Wales and Adam of Eynsham recounted the story of how the unusually large swan suddenly appeared at the bishop's manor of Stow, near Lincoln, on the day of Hugh's installation as bishop. The bird displayed striking affection for Bishop Hugh when he visited, and appeared to know when the bishop arrived in Lincoln and when he suffered his final illness and death.[53] Such ideas were given formal recognition in richly illuminated bestiaries, which collected and presented information on marvellous creatures, plants and stones. Folios 35r to 36r of Ashmole 304 are part of the *Predictions of Socrates*, with this group offering 'Spheres of Flying Creatures', reflecting the idea that flying creatures could pick up planetary influences that were active in the air.

The procedure is as follows: the Enquirer locates the theme in the opening table; is then directed via the 'spheres' of flowers, birds or renowned cities; and finally receives an answer from one of sixteen Kings. These, like the cities, are mostly from far away or long ago, and add to the aura of prophetic mystery deriving from both the natural world and ancient wisdom.

This array of sources of knowledge, insight and guidance is impressive in itself, and goes considerably further than the brief texts found in the Eadwine Psalter. The same can be said of the presentation of the prognostic texts and techniques. The texts on

Figure 3.17. Part of the *Predictions of Socrates*. Oxford, Bodleian Library, Ashmole MS 304, f35r.

palmistry and onomancy are additions to the psalter, and appear to be prized as part of the impressive technological advances being applied in the cathedral and priory. However, in Matthew Paris' book, the divinatory tables are very much the main focus. What links the Eadwine Psalter to Matthew's work – and links their divinatory texts in particular – is their liminal position between cloister and court. In both cases, powerful and respected monastic communities appear to be seeking out acceptable sources of supernatural insight into the future – and to be advising highly placed patrons as a result. The fact that the thirteenth-century book goes much further in its celebration of such insights is paralleled by contemporary developments elsewhere in Europe. As the following chapters will show, the rulers of Iberian kingdoms (and contemporary emperors and even popes) were all seeking texts and experts able to offer magical guidance and power.

4

ASTRAL MAGIC IN THE COURT OF ALFONSO X

The collection of magical theories and instructions known by the mysterious title, *Picatrix*, was originally brought together in the middle of the tenth century in Arabic al-Andalus. It draws widely on teachings from across the Islamicate world, and especially on new work done in the fields of astronomy, astrology, medicine and natural philosophy. It moved across political and cultural borders and was translated from Arabic into Castilian, Hebrew and Latin.[1] The likely author, Abū 'l-Qāsim Maslama ibn al-Qāsim Ibrāhīm al-Qurṭubī al-Zayyāt, was (disparagingly) described by a biographer as expert in charms and talismans.[2] The book itself unashamedly sets out to cover both the theory and practice of learned magic, and was warmly received by King Alfonso X, ruler of Castile, Leon and Galicia and aspirant ruler of the Holy Roman Empire. Alfonso had it translated from Arabic into Castilian (c.1256) to increase its availability. Ongoing demand is shown by the fact that Alfonso's version was subsequently translated into Latin, most likely in Italy.[3] Both versions give full and often challenging details of magical procedures, as this chapter will show.

In Alfonso (known as 'the Learned') this magic found an ideal patron, since he combined the roles of mage and wise ruler – or at least created the reputation of having done so. Alfonso's ambitions encompassed an attempt to become ruler of the Holy Roman Empire, supported by suggestions that he was a new King Solomon. Conquests by his predecessors made Alfonso heir to a large set of territories, with large Muslim populations, as well as significant Jewish communities. Much of Alfonso's cultural patronage and political scheming can be seen as attempts to unify, or at least satisfy, important groups within this diverse and complex realm. An ability to derive guidance, and even power, from celestial bodies would be even more attractive to him than to the ruler of a less challenging territory. His political ambitions were never fully realised, and indeed led to a turbulent reign.[4] Nevertheless, Alfonso made his mark on the wider European stage; and his cultural programme contributed to his reputation. Part of this programme was the collection of an impressive body of astronomical, astrological, mystical and magical works.[5] Unfortunately, not all of the Castilian versions of these works survive in full, and the lists given in known manuscripts are not entirely clear. However, enough survives to suggest that Alfonso's title of El Sabio could be translated as 'the Magus'. It is also likely, as we see below, that Alfonso intended his magical books to be put to use by professionals working at his court.

The works collected, edited and translated cover an impressive range of magical skills and technologies. They included a wide-ranging lapidary (or 'book of stones'), which provided details of the powers and astrological affiliations of a long list of precious and semi-precious stones. Even Alfonso's *Book of Games* included instructions for astrological competitions, alongside guidance on how to play chess. The collection of magical works included treatises on judicial astrology and 'Astromagia', or astral magic, as well as a survey of astrology under the title *Libro del saber de astrologia* (*Book of the Wisdom of Astrology*). All this might appear to contradict Alfonso's claim to be a faithful Christian, but his own views are made clear in his law code (the Siete Partidas).[6] This states that divination by the stars may lawfully be carried out by masters of the art, since their judgements are based on scientific knowledge. All other forms of divination are forbidden, along with necromancy, although it is recognised that even forbidden magic may be performed with good intentions.

PICATRIX AND THE FULLY TRAINED MAGICIAN

The book *Picatrix* brings together a wide range of texts and procedures, linked by the authorial figure of the 'philosopher' whose name it bears. It addresses readers willing to follow complex, technical instructions in order to contact and control spirits. Knowledge of the occult powers present in earthly substances, such as stones and plants, is to be combined with the ability to use images and talismans and to channel the powerful rays emitted by celestial bodies. The latter requires understanding of the factors affecting the strengths and interactions of the emitting bodies, as well as the power to manipulate the spirits who act as intermediaries. For Christian moralists, the invocation of such powerful spirits would be at least troubling, if not actually heretical or diabolical. However, in the world view of the text, the intellectual and moral force of the philosopher-practitioner means that he (and it is always a matter of 'he') dominates the spirits, rather than being tricked by them or entering into pacts with them. Moreover, the powers of the spirits, like those of the celestial bodies with which they are in contact, are part of God's creation. It follows that they are capable of being grasped and controlled by a human being who is sufficiently strong and prepared – and this is claimed to be morally acceptable, since the human controls the spirit. In other words, this is magic for the professionals who would devote years of their lives to it.

That is important since the translations and editions compiled for Alfonso suggest that the king expected a return on his investment. The magical and supernatural knowledge is presented as contributing to the power and superior status of the king. The collection entitled *Libro de las cruzes* was translated in c.1259 and explicitly linked royal policy to understanding of the 'science of the stars' and of astrological judgements. Alfonso was here acclaimed as a new Solomon, and his judgements were to be equally valued. The *Bocados de oro* (*Nuggets of Gold*), translated in c.1260, further reinforced this by presenting advice for kings from the mage Hermes. More mystical, but equally magical, was the edited work usually known by its Latin title of *Liber Razielis* (*Book of Raziel*). This drew upon Jewish, rather than Islamicate, works of magic, while claiming both Solomon and Hermes as sources.[7] The *Book of Raziel* claimed to transmit secret knowledge of divine origin, brought to humans by the angel Raziel. The Latin version has a prologue stating that Alfonso had a team of Masters helping with his collection and 'improvement' of the works of philosophers and sages. It also explains the title of the work, as well as referring to Solomon.[8]

Sadly, Alfonso's own copy (or copies) of *Picatrix*, whether in Castilian or Latin, does not survive.[9] However, part of the work on astral magic, the *Astromagia*, has been identified in Vatican, Biblioteca Apostolica, Cod. Reg. Lat. 1283a (folios 1–36); and what may be draft versions of astrological works from Alfonso's collection are found in Madrid, Biblioteca Nacional, MSS 3065 and 9294.[10] Tantalisingly, Madrid, El Escorial, Real Biblioteca, MS h I 16 gives a summary of eleven treatises on the subject of heavenly images, their powers and the powers of certain stones.[11] This gives the date 1269, together with the statement that the collection was compiled on Alfonso's orders from the works of ancient philosophers, and the title – *Book of the Forms and Images Which Are in the Heavens*.[12] No details or illustrations of the images themselves are given.[13]

Alfonso's collection certainly included works detailing magical images whose powers derived from the heavens. The history of his books can be pieced together, and shows that other European rulers sought treasures from Alfonso's library. The famous bibliophile, the duc de Berry, brother of King Charles V of France, commissioned a French translation of a work described as 'the Spanish Book of forms, figures and images which are in the heavens', and Charles himself had a copy of the French version in his own library. This copy was presumably among the volumes on astronomy and astrology seized by John, duke of Bedford, in 1425 and sent to England in 1429. Its fate thereafter is unknown. Alfonso's book of treatises on astral magic seems to have been compiled later than the *Book of Forms and Images*, since it refers back to that work. What may have been Alfonso's own copy of this *Libro de astromagia* (*Book of Astral Magic*) – or a close copy of that original – appears in the 1373 catalogue of Charles V's library. It is described as 'a book of astronomy, possibly the *Ars notoria*, in Spanish, finely decorated with figures and painted illuminations in the Bolognese style'.[14] This codex appears not to have been taken by the duke of Bedford, since the catalogue description accords with the fragment now included in Vatican, Biblioteca Apostolica, Cod. Reg. Lat. 1283a. That complex volume came into the hands of Queen Christina of Sweden, whose books were transferred to the Vatican during the papacy of Alexander VIII (1689–91). For that reason, the images in this fragment may be taken as the closest now surviving to the rich sets of images described in Alfonso's collection of magical works, including *Picatrix*. However, in order to set the scene, it is important to look first at the surviving text of *Picatrix* itself.

What the book offers, through the elusive persona of Picatrix, its supposed author, is a wide-ranging collection of practical instructions for specific rituals with specific outcomes, presented under the umbrella of a scholarly discourse on the powers that make the magic and the rituals work. There is repeated stress on the idea that only very special individuals, equipped with a high level of both education and personal strength, can successfully apply this magic – while the unlearned and 'unwise' will be safely repelled by the complexity of the system of ideas deployed and the technicality of the language. The appeal of all this for those aspiring to the status of mage is obvious, as also is the commercial potential of the outcomes on offer.

As would be expected of a learned work, *Picatrix* opens with a prologue, setting out the credentials and coverage of what is to come. And as was customary for works offering induction into the practice of magic, this prologue promises knowledge both powerful and secret. The condition is that time and effort must be devoted to philosophical and magical study. Knowledge can be obtained by following the programme laid out in the book, but of course it will not be easy. Picatrix explains that the philosophers used mysterious words, images and symbols. This is because their knowledge is so powerful that in the wrong hands it could destroy the whole world. The learned Picatrix is already an expert, and can offer guidance in both the true understanding and the practical applications of this dangerous material. Even so, the work must be kept away from anyone who might misuse it. The prologue then ends with an outline of the course of learning that is promised: Part 1 sets out the nature of the heavens and the powers of the celestial signs; Part 2 expounds the spheres of the universe, with their powers and earthly effects; Part 3 covers the planets and the zodiac signs, providing details of the figures, forms and colours associated with them, and of the spirits of the planets; and finally, Part 4 includes practical insights into the powers of the spirits and how the practitioner can make use of such materials as images and suffumigations.[15]

The first chapter explains that all created things (including living beings) have natures and properties assigned to them by God. It also repeats the common medieval view that God organised all of creation hierarchically. The promise of the book is that it will provide insight into the workings of God's creation and all its component parts. Penetration into this divine knowledge and its effects is the first of the secrets to be attained by the student of magical wisdom. Without delay, the next chapter is headed 'What Magic Is and What Its Properties Are'. Still more enticingly, magic is defined as acts performed by those able to apply both their spirit and their senses to the creation of effects so wonderful that they appear miraculous. Indeed, because magic can draw upon divine secrets, it is beyond the grasp of the ordinary intellect, as well as the senses.

More practically, the expert is able to operate with spirits, images and matter, in order to bring about transformations in the appearance or qualities of things. These are not mere sensory illusions, but they can even defy the essential natures of the chosen substances and creatures. Such powers cannot come simply from human resources, and accordingly readers are assured that they will be given the knowledge required to construct talismans of great power.[16] The talismans will combine substances, images, mathematical harmonies, celestial influences and spiritual forces in such a way that – if made and used at the correct times – they can attract appropriate spirits and call upon their powers. What is required to make use of this power is knowledge of the appropriate words. When all these things are brought together at the correct hour and thus under the correct alignment of the celestial bodies, then the power of the magic will be released. This magical power then needs to be controlled and utilised. For this to work, heat and appropriate suffumigations are also required; but the spiritual force of humans and animals must also be drawn upon. These are harnessed through 'works, deeds and other subtle means'. Mages able to do all this will, the text promises, have achieved a state of exalted knowledge and privileged understanding of God's universe. They will also have subjected themselves, and others, to extreme experiences.

The text moves on to provide the promised guide to the structure and powers of the heavenly spheres. The author writes as an expert, able to pick key points from a wide range of authoritative sources and to highlight the errors of some assertions and ideas. Chapter 3 provides a neat summary of the system:

The nature of the heavens is one and singular, and all motion [below], whether of bodies or natural forces, is propelled by the motion of the heavens. All heat flows down from the heavens and so all of Earth's heat comes also from the heavens. The heavens are divided into three hundred and sixty major degrees, and the number of their figures is the same. Through these all astrological judgements are reached, since astrologers draw upon the images of the heavens, and all change below is caused by the heavens.[17]

Those wishing to make an astrological judgement need to know the qualities and powers of each degree, together with the qualities and powers of each planet both in general and as inflected by passing through each degree of the zodiac. Moreover, they must also know the ways in which the planets affect one another as they move at different speeds (and sometimes in different directions) around the 360° of the zodiac belt. For example, planets very close to one another are in Conjunction, whereas those facing one another across the zodiac are in Opposition. If planets are at 30°, 60°, 90° and 120° to one another or to the points at which the Moon's orbit intersects with the ecliptic, this will add further significance. The latter were frequently referred to as the head and the tail of the dragon and were treated as if they were planetary bodies.

The next section, which begins with the exposition of the making of talismans and images, is more advanced. It emphasises the powerful effects of the constellations outside the zodiac, as well as the complications caused by the fact that the planets were believed to be set in spheres, which were themselves rotating around the Earth. More usual is the insistence that no trials or half-hearted attempts at this magic will ever be successful. Equally, the talismans themselves must be carefully and skilfully made, or they, too, might cause failure. More positively, the text moves on to a detailed exposition of the mansions of the Moon, claimed to be those unanimously agreed upon by all the sages of India.[18] What follows is a list of the twenty-eight portions into which the belt of the zodiac is to be divided, beginning with the first degree of Aries. For each mansion a name is given, often related to – or modelled on – the Arabic names for major stars. No information on the qualities of the mansions themselves is included; instead, interest is focused entirely on the functions of talismans and images made when the Moon is in each mansion. For instance, the fourth mansion is Aldebaran, and runs from the eighth to the twenty-first degree of Taurus. When the Moon is here, images with many negative applications are to be made – to destroy villages or cities, to cause hatred and to poison wells. By contrast, the seventh mansion is Aldirah, and runs from the seventeenth degree of Gemini to the end of that sign. Images made under the influence of this mansion can increase harvests and trade, protect travellers, strengthen relationships and help gain favours from the powerful. At this point no explanation of these powers is given.

The practitioner is expected to be expert in astronomy and arithmetic, since most of the lunar mansions have boundaries specified to minutes and seconds of a degree, rather than to a whole degree. Moreover, information on the Moon's aspects with other planets is also necessary. If an image to produce positive outcomes is to be made, then the Moon must be not only in an appropriate mansion, but also in beneficial aspects with the fortunate planets. In other words, the Moon should be in trine (120°) or sextile (60°) aspect with Jupiter and/or Venus. Equally, the Moon must be free from any aspect with the harmful planets – Saturn and Mars – and from 'combustion' by the Sun (i.e. it must not be too close to the Sun). For talismans to cause harm, the reverse applies. This is potentially challenging, since the Moon moves through approximately 12° of the zodiac every twenty-four hours, and is thus in each mansion for at most one day. If the Moon is subject to an inappropriate aspect at that time, then a significant delay could be caused in making the desired talisman. Sadly, even if all the factors are propitious, problems can still be caused if the human practitioner is not in a sufficient state of harmony with the powers that are to be utilised, or if a poor choice of material is made for the talisman itself.

The choice of time for using a talisman is also significantly affected by the Moon. If the ritual is to take place during the hours of light, then the Moon must be ascending and in a diurnal sign. The latter term may signify either a sign considered to have masculine characteristics or one placed above the eastern horizon. For night-time rituals, the Moon should be in a nocturnal sign – though again ascending. Still more restrictions are caused by the time taken for the Moon's sign to rise. Signs of direct (and thus longer) ascent are here considered more propitious. For this purpose, these are stated to be the six signs from Capricorn to Gemini; the signs from Cancer to Sagittarius are thus those of indirect ascent. If the Moon is in an indirect sign, then the power of the image will be weaker and slower to take effect. However, if the need is urgent and the magician cannot wait very long, it will be sufficient to ensure that Jupiter or Venus is on the ascendant or at midheaven.

The next section, 'Examples of Cases for Discussion', turns to a discussion of complex theoretical issues. For example, the second 'case' concerns a technique for bringing about love between two people. The outline of the procedure itself is pared down to the essentials, and simply says that two images should be made, one of each person, 'under the ascendant of the querent's question'. This implies that the magician is likely to be working for someone who has come to make a request (the querent). The rest of the instructions are terse, saying: 'let the ascendant and the tenth house be fortunate, with the ascendant safe from the influence of the unfortunate planets. The lord of the tenth house should be fortunate, and facing the lord of the ascendant in a trine or sextile aspect.' What follows is, in fact, a discussion of the astrological concept of aspects, and why some are suited to specific applications (in this case, love magic). The user is also expected to be familiar with the 'faces' (divisions of a zodiac sign, corresponding roughly to ten days) or subsections of the signs, and with Hindu-Arabic numerals (still relatively novel in western Europe). However, it is only the making of the talisman that needs this complex range of knowledge and skills. For use, what matters is simply physical proximity to the person(s) or places to be affected, and customers could presumably handle that part of the procedure.

This set of instructions is addressed to a reader familiar with the concept of astrological houses, as well as of zodiac signs. Medieval Latin texts are often unhelpful in referring to zodiac signs as houses, but there is an important distinction between the two. As already stated, zodiac signs are effectively sections of a belt centred on the ecliptic (itself the path apparently followed annually around the Earth by the Sun). The zodiac is referred to as a 'belt' because it extends several degrees above and below the Sun's path. The zodiac, with the ecliptic at its centre, appears to rotate around the Earth every twenty-four hours, carrying the zodiac signs and the planets with it. By contrast, the houses are calculated in relation to the Earth itself and have the eastern and western horizons for any given location as key points. Therefore, they do not move; and their significance is specifically related to parts of human life on Earth. A key point for these 'terrestrial' houses is the Ascendant, located on the eastern horizon and forming the starting point of the first house. As the heavens rotate around the Earth, the whole zodiac will pass through the Ascendant every twenty-four hours. A planet or sign on or near to the Ascendant at any chosen time is in a powerful position and can give a key to the reading of a complete horoscope.

The final concept that the would-be practitioner of this magic needs to grasp is that of the planetary days and hours. These portions of time are frequently referred to in the instructions for preparing images, alongside the signs and houses that allocate portions of celestial space. For instance, still in the section on 'Examples on Cases for Discussion', the following guidance is given on making talismans to gain love, affection or favour:

make the talisman on the Moon's day [Monday] with the Moon full and in Sagittarius, Taurus, Cancer or Pisces. If the Moon can be at the Head of the Dragon that will increase the talisman's power. The Moon must also be in a fortunate house which is appropriate for the purpose. Let the Moon be linked or close to Jupiter in the hour of Jupiter, and with Jupiter placed in Pisces, Sagittarius or Cancer.

Each planet (including the Sun and the Moon) was assigned a day, with Sunday being the Sun's day and Monday that of the Moon. Tuesday is the day of Mars, Wednesday that of Mercury, Thursday that of Jupiter, Friday that of Venus and Saturday that of Saturn. The first hour of each day is assigned to the planet ruling that day, and this first hour is frequently (though not always) taken as the first hour of daylight. This lack of certainty was linked to the different systems for calculating a day which were found in Latin, Hebrew and Arabic sources. Bede had noted that the biblical book of Genesis stated that 'the evening and the morning were the first day' and that Mosaic feasts began at sundown, whereas the Romans started their days at midnight, the 'Chaldeans' at dawn and the Egyptians at sunset.[19] The subsequent hours are assigned one to each planet, in a sequence following the perceived order of the planets from outermost (Saturn) to closest to Earth (the Moon). On a Monday, the hours of Jupiter would therefore be the third, tenth and seventeenth, with the last being a nocturnal hour.

It should be noted that the author pauses at this point to assert once more that absolute faith is essential, along with complete concentration and total secrecy. Maxims claimed to be from both Plato and Aristotle are called upon to support this. Aristotle's expertise in magic is presented as having been so great that he knew the names of spirits of enormous power, as well as how they could be called down and actually into the body of the magician. This is highly dangerous for the uninitiated; but Picatrix makes it clear that he is passing on tested knowledge and advice. He has already provided a list of names, with instructions on how they are to be copied out and how the seal of Solomon is to be placed near them. Picatrix presents himself as able to use such names safely; but he is careful to stress that words, even when coupled with prayers, have power over spirits only when granted by God.

This introduces the next level of magical knowledge to be discussed, namely that of the use of words when performing magic. Powerful words may be incorporated into the talismans themselves, recited when making them, and uttered when using them. If the talismans draw upon and encapsulate planetary forces and powers, then words clarify the exact outcome desired when a talisman is put into action in a ritual. Words also express the power, faith and will of the magician in relation to both celestial forces and individual spirits. Accordingly, a verbal formula is now given to be spoken when using the talismanic images of the couple who are to love one another. In this formula it is the spirits of the man and the woman that are to be affected by the magic, not just their bodies. Examples of formulae of similar types are also provided for a range of purposes, negative and positive. Some offer defence against magic, or removal of a hostile spell. If it were not already obvious, this makes it clear that this is both a theoretical discourse on how magic can work and a guide to using magic in practice. However, the subject for the moment is magical theory, and the text accordingly returns to the power and importance of such knowledge.

Only the wisest humans, we learn, are able to understand the next part of the book. This deals with the definitions and interactions of the First Principle, matter, the elements and perceptible forms, as well as the power relationships between intellect, spirit, bodily forms and human individuals. The system is complex and fluid, but informed human intellect is capable of grasping the necessary powers and of harnessing the appropriate words to call upon spirits. Thus, magical rituals of sufficient complexity are capable of

bringing about real effects. Moreover, matter and form, while fundamental to physical reality, are subject both to natural processes of
change and to human intervention. With sufficient learning, meditation and effort, the magician will even become capable of understanding the words granted by God to Adam, with all the power
invested in them. The attraction of such superhuman knowledge
and power, at least to those with the will and resources to devote
their lives to the pursuit, is clear. Equally, those who wished to attain
or retain wealth and power would find it useful to be able to call
upon the services of such great mages. The extraordinary abilities of
the experienced mage emerge still more powerfully in the 'practical'
section of the book, which expounds some of the wonders they will
be able to perform.

Meanwhile, Book Two moves on to a more detailed survey of the
key elements of astrology, relating this to the experiences and studies
of Picatrix himself. He started by studying the work of Ptolemy of
Alexandria, before moving on to Egyptian and Indian techniques.
The Indian astrologer had a collection of talismans or 'sigils' (magical
characters), each of which apparently gave off a noticeable, incense-
like perfume and could be administered in a drink. One of these,
which especially impressed Picatrix, cured a man who had been
stung by a scorpion. It carried an image of a scorpion, impressed
into the material of the sigil by a carved stone, itself set in a golden
ring. This talisman stone had been engraved while the Moon was in
the second 'face' of Scorpio (the sign of the scorpion). Inspired by
this, Picatrix made a similar image, which worked at least as powerfully as that of the Indian astrologer.

Disappointingly, Picatrix records that his attempt to expound
the magical figures of the zodiac in relation to geometrical and
astronomical data has been hampered by the obscurity of the
language used by earlier sages. He explains that this obscurity is
partly caused by the fact that the constellations – including those of
the zodiac – move slowly in relation to the divisions of the ecliptic

(this is the phenomenon now known as precession). For that reason, they were no longer to be found in the positions observed by ancient astronomers, or even by Ptolemy. By contrast, the signs, when considered as portions of the ecliptic, remain unchanged, and so also, according to Picatrix, do the figures or images discerned by Indian astrologers in parts of each sign. Picatrix is clear that these 'Indian' figures are accurate and powerful, and that their power can be drawn upon by those with sufficient knowledge.[20]

As examples, the figures for the three 'faces' or divisions of Aries are described in detail. In the first is a man with red eyes and a long beard, wrapped in white garments with a rope belt, and striding powerfully (with one foot raised) while looking at something in front of him. In the second is a red-clothed, horse-faced and one-footed woman, in search of clothes, jewellery and a son. The third and final figure for Aries is a man who is both red and white, having a sword in one hand and a staff in the other, and is expert at working with iron. Picatrix states that the Indians derived these figures entirely from the natures of the signs and stars. His appeal to the imagination is a warning that readers should not expect to see actual figures in the night sky. That Aries was said to be ruled by Mars (whose metal was iron), to be the 'Exaltation' of the Sun (i.e. the Sun was at its most powerful when in Aries) and to be masculine and fiery would have been known to most readers. The assertion that the figures were Indian in origin suggests that the author/editor of *Picatrix* has used relatively recent sources, since the images can in fact be traced back to Egyptian concepts which were later translated into Pahlavi and used by the great astrologer, Abū MaʿShar.[21]

This promisingly precise information is followed by injunctions that the would-be mage must work through a long programme of study into the exact effects of each degree of the zodiac when occupied by each of the planets, and with each possible combination of planetary aspects. The sum total of these is said to be 10,080. This will create the expertise required for actual understanding of the making and use of the figures. The next chapter moves on to the next layer of necessary information, which is that of the cyclical interactions of the Sun and the Moon through each lunar month, and how these are inflected by the zodiac signs occupied by each. Clearly, it is only after long preparation that the magician is ready to perform the promised magic. Even then, great care must go into choosing the best time and place.

First to be considered are the path and aspects of the Moon, together with the positions of planets positively associated with the Moon's sign. It is necessary for the planetary 'ruler' of the Moon's sign to be in a positive position and near to the ascendant. This planet must also not be retrograde (moving backwards in relation to the ecliptic). The same should be done for the Sun; and care should be taken that neither Mars nor Saturn is on the ascendant or at midheaven. Further power is provided if the ruler of the ascendant sign is close to the eastern horizon, and if the fortunate planets (Venus and Jupiter) are well placed. In fact, Picatrix provides lists of desirable factors for any given outcome that are so challenging as to be very difficult to bring together on demand. However, practical solutions are offered. For instance, if no positive aspect between the Moon and Venus is available for a procedure relating to love or friendship, then a positive aspect between the Moon and Jupiter will also work. Failing that, a positive aspect between the Moon and the ruler of the sign occupied by Jupiter will be sufficient. Failing all that, having the Moon near Venus, well placed in relation to Jupiter, and free from negative aspects will do. Similarly full and technical guidance is given for a wide range of possible goals. However, nothing has yet been said on how to bring the astrological requirements together with the appropriate figures and words in practice.

This comes a little closer in Chapter 6 of Book Two, which states that magical images work through having 'virtues' which link them to the powers of celestial bodies. These create a flow of celestial power under the control of the magician. The first step is the making of the image. Material, form and time are all crucial if the appropriate spirit is to be drawn into the image. Also needed, though discussed in obscure terms, are materials from living beings. The text suggests that magicians will gain power over living creatures – and that this is just as possible as the brewing of beer or the making of cheese. Thus, magic is also an elevated form of technology. Special emphasis is placed on understanding the effects of the motion of the sphere of the fixed stars, together with the powers of those stars. Stress is placed on the fact that the sphere of fixed stars moves all the other spheres below it. The concept of the *primum mobile* or First Mover is invoked to support this. From this point, the text shifts more directly towards instruction in actual magical rituals.

It is in Chapter 9 of Book Two that selected practical examples are given, in the form of six images. The first is for driving out mice, the second for repelling insects and the third for driving away flies. The fourth will attract a chosen person, the fifth will create hostility between people and the sixth will render a place uninhabitable. Each is relatively simple to make, as an example will show. The fly repellent is to be made when the third 'face' of Scorpio is ascending, and requires three magical characters to be inscribed on a thin sheet of tin. The characters (which are illustrated in the text) provide the link to the power of the fixed stars. The tin provides the link to the benevolent planet, Jupiter, while the timing must be chosen to create a connection to Scorpio. The book then proceeds to deal with the more complex procedures required to produce many other types of image. This practical information covers the stones, metals, figures and images associated with each planet. Instructions are also given for making powerful rings. By now it will come as no surprise to learn that precise astrological timing is required.

For example, to make an image of Jupiter that will bring lifelong success and good fortune, it is necessary to acquire a clear, white stone of a type ruled by the planet, and to calculate the hour of

Jupiter on a day when Jupiter is both ascending and in his own zodiacal Exaltation (Cancer). This would be no spur-of-the moment thing, since Jupiter moves slowly around the zodiac and remains in each sign for approximately a year. The image to be engraved is also demanding. Jupiter is to be shown as a crowned king, speaking, and seated on a throne whose feet rest on the necks of four winged men.

Some of the images also utilise characters (of the types specified for the six images described above). A relatively simple one is linked to the Moon. A large piece of crystal is needed, as is an hour of the Moon that coincides with the Moon itself being close to the Ascendant. On one part of the crystal must be engraved the figure of a woman with tousled hair, and with her feet on the necks of two bulls. Another part should show a standing woman wearing a crown and holding a cane in her right hand. A sequence of four characters is to be placed around her, one of them being a reversed crescent moon. If this image is impressed into wax and the wax placed in a dovecote, then a good supply of doves will be ensured.

Of course, the signs and stars, as well as the planets, have powers that the magician needs to harness, and it is therefore logical for Picatrix to move on to a fuller account of the three 'faces' of each zodiac sign. What is surprising is that the descriptions given here, although also attributed to the 'sages of India', are different from the earlier ones. Each face is here linked to a specified planet, beginning with Aries and Mars. Plato is claimed as a master of such magic, as is Pythagoras. This provides the resounding conclusion to Book Two.

Book Three offers still more information on the planets, signs and figures, but adds the inducement of teaching the would-be mage how to speak to planetary spirits and thus progress further in magic. More details are given here of the medicinal properties and physical effects of the planets and zodiac signs, via the body parts which they rule. The mage is now finally ready to move on to the construction and performance of complex rituals for harnessing the powers of the planets. Clothes of appropriate colours and fabrics should be worn, such as black wool for Saturn, pink silk for Venus and bright white silk or linen for the Moon. The air must be infused with appropriate suffumigations, such as pepper for Mars and musk for the Sun. Words or images must be written or painted in the correct colours, prepared in the manner stipulated. For instance, the colour of the first face of Virgo is red fading to gold, and this is to be made by mixing powdered saffron into green gall in water, and then adding a small quantity of gum. The true mage will also eat and drink only those things attuned to the desired celestial powers, and will know the appropriate prayers and sacrifices.

That last point makes it clear that Christian theologians would accuse practitioners of idolatry and heresy. However, in Chapter 4, Picatrix goes on to reassert that only the truly wise and learned will be able to use all this mystical knowledge. The fully prepared magician, working at the correct time and with the correct figures, will select substances from animals, plants and minerals and combine them into foods, clothing, incenses and suffumigations. The process of burning is important since it uses the power of fire to release these compounds into the air and thus link them more directly to the realm of the stars. The qualities of the air are also important, since air is described as a body which receives forms, influences and planetary effects. The living breath of the magician will bring the suffumigation, the air and the human body into dynamic contact. With all this in place, the magician should be able to make contact with the chosen planetary spirit, although such spirits cannot be perceived directly. The magician should also prepare the necessary vessels, in the forms of a cross and a censer (of specified shape and material). All these things are to be used in a building reserved for the purpose and with an opening in its roof. The smoke of the suffumigation should flow from the censer through the cross and through the magician's body, before flowing out through the opening and up to reach the chosen planet. If the correct hour has been calculated, then no unwanted planet's rays will be in the way of the smoke. This will establish contact with the planetary spirit. Consequently, the rays of the planet, with the spirit, will flow to the magician and produce magical effects. Nevertheless, the process is difficult and prone to failure, and so further instructions are given.

The first is the revelation that Aristotle and other sages identified four spirits of 'Perfect Nature' and taught chosen pupils their names. These are now revealed by Picatrix, and confirmed by a long narrative linking Hermes, Aristotle and Socrates to the fourfold spirit of 'Perfect Nature'. There follow precise instructions for rituals to make contact with the spirits of each planet. For instance, Saturn is to be addressed in his hour and on his day (Saturday), and when he is in a sign that increases his power (Libra, Aquarius or Capricorn), with care being taken that the planet should not be in a position that will bring out destructive effects, or in any negative aspect. It is crucial that Saturn should not be retrograde. The magician should be dressed entirely in black, topped with a black academic robe, and should hold an iron ring. The censer should be iron and the suffumigation should include opium, laurel seed, wormwood and the head of a black cat. While the suffumigation burns, the mage should look towards Saturn and recite a prayer, which is given in the text. Fortunately, the suffumigations for the more benevolent planets are less alarming, although the practitioner is required to eat such things as the livers of cats, mice and cockerels.

The book moves on to provide detailed prescriptions for rituals to attract multiple spirits for each planet, starting with their names. Once again, all this is attributed to Aristotle, who supposedly taught how to make four compound magical stones. These stones involve powdered diamonds and rubies, gold, silver and mandrake, as well as arsenic, yellow sulphur and wolfsbane. The dangers of working with such poisons are emphasised, and practitioners are given recipes for antidotes, as well as instructions to cover their faces, plug their ears and protect their hands with a special medicine. Body parts from lions, leopards, wolves, pigs, crows, sparrows and horses are all utilised, as is human blood. Human sacrifice is not required, but a section of Book Three is entirely dedicated to magical uses for parts of the human body. The procedures are long and unpleasant, although the humans are already dead when their body parts are taken. The desired effects are also negative, producing the apparent transformation of humans into animals or monsters, or loss of memory and sanity. Bodily substances, especially dried human excrement, human urine and menstrual fluid, are asserted to have powerful medicinal effects, which are listed. Also terrifying are the recipes for drugs to induce sleep or immobility: these involve mandrake roots, poppyseeds, opium, henbane, datura, arsenic and wolfsbane, as well as animal parts and various liquids. By this stage, the magic is in danger of descending from its claimed mystical heights; but Book Three reassuringly concludes with a reiteration of the high principles to be followed by the magician. This reads like a conclusion to the whole work.

However, there is in fact a Book Four in the Latin version of *Picatrix*, which has the appearance of a set of appendices. Egyptian and 'Chaldean' experts feature more strongly here than in previous books. It opens with a philosophical discussion of the structure of the created universe and the key, intermediate position of the 'circle of spirit and spirits'. This is followed by an outline of instructions from Chaldean and Egyptian sages on the invocations and roles of the Moon's spirit as the planet moves through the signs of the zodiac. After these instructions, the text moves abruptly on to a further set of suffumigations, claiming Indian sources and using magical characters or sigils. Chapter 3 expounds on the Chaldean experts or 'magi', while Chapter 4 offers a selection of sayings, claimed to be from the *Secret of Secrets*, but more technically magical than most versions of that work.[22] Excerpts are also given from the *Centiloquium* attributed to Ptolemy and from a text referred to as the *Chaldean Agriculture*.[23] As the author/compiler notes, all these materials add to the knowledge required for the successful performance of the rituals. They also emphasise the credentials of the book as a compendium of learned magic from glamorous realms distant in place and time.

The final portion of Book Four is still more mixed, and little effort is made to integrate it with the preceding texts. It is not found in the Arabic version of *Picatrix*.[24] It promises 'composite' images, capable of producing 'miraculous works' – and yet the text begins with a sequence of 'receipts' for making compound substances to repel pests, cure insect bites, catch fish and birds, perform cures or deceive the senses of watchers. The making of images appears only at receipts 14 and 15, with practical instructions for making a bronze image of a monkey and filling it with scammony for use as a contraceptive, and for making an image of a dog from mandrake root to scare away dogs. However, sections 16–18 have overtones of alchemical procedures in their instructions for distilling 'vinegars' which can turn iron into water or make water burn. There is no attempt at a conclusion, simply the statement: 'Thus ends the book of the wise Picatrix.'

PRACTISING ASTRAL MAGIC IN THE COURT OF ALFONSO X

From the above, it is clear that aspiring magicians would need to spend a considerable period of time following the course of study set out by Picatrix. They would also need precise calculations of planetary positions in relation to the zodiac and to one another. For those working in the court of Alfonso X, it was very helpful that the king commissioned two scholars to produce an updated version of the existing astronomical tables. The experts concerned were two Jewish astronomers, named Isaac ben Sid and Judah ben Moses ha-Cohen.[25] The prologue states that the tables were a correction of those compiled by 'Azarquiel' – now usually known as the Tables of Toledo – in the late eleventh century.[26] Careful attention was paid to the stars, as well as to all seven planets; and a list of 1,028 stars, based on that of Ptolemy, was appended to the planetary tables, giving their positions at the time of Alfonso's reign.

However, a reading of Picatrix's instructions for the performance of magical rituals shows that magicians needed considerably more resources than updated planetary tables. These included an extremely well-stocked pharmacopeia, large quantities of precious and semi-precious stones and metals, and the services of skilled artists, metal-workers, sculptors and builders. For instance, Alfonso's composite *Lapidary* strives to identify a stone (and associated fixed star) for each of the 360° of the zodiac. In addition to all this, specially built premises and furniture are needed for some rituals. The text simply assumes that all its instructions can be carried out, although it acknowledges that failure is always possible. The scale of the work of collecting, translating, editing and illustrating astrological and magical works that was carried out in Alfonso's scriptorium shows that the king was prepared to invest substantially. It is therefore probable that he paid for the services of professional magicians striving to use the assembled knowledge. Evidence for this comes from other texts, which were once part of Alfonso's collection of specially produced magical books.

The distinctive style of the illuminations in surviving volumes suggests that a team of artists and illustrators was employed to produce carefully designed images. These in turn could provide exemplars for artisans tasked with manufacturing the complex talismans, images, censers, figures and other objects specified by the texts. In certain cases the designers had to draw upon research going beyond what was stated in the nearby text. Some illuminations even depict professional magicians in the act of carrying out the rituals. All these can be seen in the surviving portion of Alfonso's *Astromagia*, which is now Vatican, Biblioteca Apostolica, Cod. Reg. Lat. 1283a.[27] This manuscript has received surprisingly little attention, and a close analysis of it produces important findings. The illuminations provide a detailed insight into how the magician was expected to perform the magic outlined in the royal collection.

The first section of the fragment gives verbal descriptions of the 'figures' identified for each degree of each zodiac sign from Taurus to Sagittarius, before making brief comments on those born under the influence of that degree. The text is given first, and in each case is followed by a large circular diagram. These echo astrological practice in having the Ascendant (eastern horizon point) on the left-hand side. In the centre of each is the symbol of the relevant sign. Surrounding this is a pair of concentric rings, divided into thirty sections – one for each degree of the sign. The outer ring contains images of the 'figure' for each degree, while the inner one contains a brief verbal explanation. The same procedure is used for all the succeeding signs. The miniature for Gemini, on folio 2v, is a good example.

Here it can just be seen that there is a small number 1 beside the two men dressed in blue in the section at centre left. This marks the first degree of the sign and shows that the 'wheel' will rotate clockwise as each degree rises across the Ascendant. It is worth noting that this is visually effective and yet slightly contradictory in relation to the fact that no zodiac sign represented a complete astronomical or astrological 'circle': each was a sector of the large circle of the zodiac. Another striking characteristic of the diagrams in this manuscript is that there is no reference to the stars which make up the constellation of Gemini. Instead, what is important here is the information about the characteristics of each degree, almost always envisaged as a living creature. This knowledge is important for the magician, since the nature of the degree, and the inflections of its power as it moves around the heavens, will affect the planets in astrologically significant relationships to it. This in turn will affect the kind of magic that is best performed at that time. A sign of the complexity of this enterprise is that the placing of the figures in the diagram for Libra (folio 6v) is slightly wrong, in that the figure for

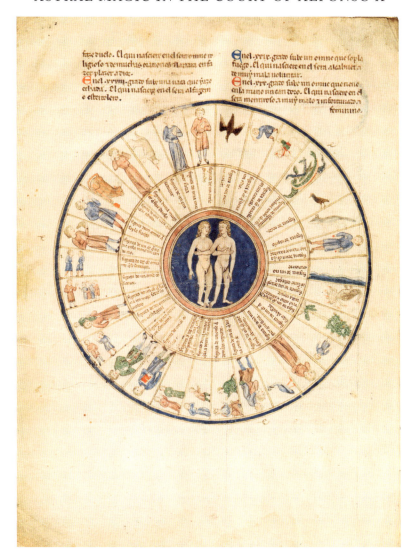

Figure 4.1. Diagram for Gemini. Vatican, Biblioteca Apostolica, Cod. Reg. Lat. 1283a, f2v.

the first degree is placed above the Ascendant. Muddles like this, together with the number of prominent corrections to both text and rubrics, suggest that this manuscript was a first attempt, rather than the final version of the work.

Magicians who wished to make use of Picatrix's material on the powers of the mansions of the Moon would find folios 13v–24v very useful. Extracts attributed to 'Kancaf the Indian', 'Plinio' (cited also in *Picatrix*), 'the Indians' (again cited in *Picatrix*) and Hermes are brought together. Here again precise and gorgeous illuminations follow the texts (although the relationship between text and image is not always clear). Each account of the lunar mansions is followed by a circular diagram, beautifully executed in the same style as the preceding miniatures. The link between text and image is so close that, where a part of the text for mansion 22 is missing, there is also a gap in the image. Skill on the part of the designers is shown particularly in the image on folio 23v, which is so complex that it provides three (or even four) tiers of figures for each mansion. Access to the other works in Alfonso's collection is assumed, as is shown in the prayers and illuminations relating to the planet Mars. The discussion of relevant images on folio 27r cites both 'our book of images' and 'Picatrix'.

Further insight into the practice of magic as envisaged in these volumes is provided by the texts and images on folios 31–36, which provide detailed instructions for magical rituals focused on the planet Mercury. The layout of the folios, in two columns, remains the same as for Mars, as does the placing of the miniatures at the end of the relevant texts. The miniatures also remain circular, although they depict events on Earth. They also continue to use a sky-blue background, reinforcing the impression that the magicians represented are achieving contacts with unearthly powers and creatures. We begin with general invocations addressed to the planet, which are followed by the recipe for a general suffumigation. These are followed, beginning at the bottom of folio 31r, by prayers and suffumigations for use when Mercury is placed in each zodiac sign in turn. The illustrations for the general texts simply show the magician standing over a brazier from which flames emerge, and raising his hands to the sky; but those that follow are much more specific. The images for Mercury in Aries and Mercury in Taurus, on folio

31v, show the magician at work. He kneels or sits on the ground, in an appropriate location, between his braziers and the animal he has sacrificed; and it is assumed that he has sufficiently prepared himself. In both images he addresses the 'figure' of the planet, shown as a young human figure riding a peacock; but this figure remains in a separate pictorial space and with its back to the magician.

Encouraging progress is shown in the following images. The fourth in the series, for Mercury in Cancer, on folio 32r, has a small, ghostly, spirit in the form of an angel, which is in the same space as the magician and conversing with him. Further successes are shown in the miniatures for Mercury in Leo and Virgo. By the time Mercury is in Libra, the miniature on folio 33r shows physical contact being achieved through a magic ring, while the ostrich eggs required for the ritual glow above the incense burner. When Mercury reaches Capricorn, on folio 33v, a bush near the magician is burning and a lion appears close by, while equipment in the form of a glowing square is hung above. The last image in this sequence, for Mercury in Pisces, marks a moment of triumph. Here the peacock-riding spirit has turned round, is facing the magician and is listening to his prayer and requests.

The suggestion that these images show the successful practice of the magic is supported by the next section of text. This goes on to explain how to make and use rings in rituals addressed to Mercury. As noted above, one such has already appeared, and been shown to be effective, in a miniature for Mercury in Libra. What is now added is that special symbols or characters should be included when making the rings, and these are provided, together with cross-references to the prayers. The powers which the first ring can bestow are listed, and this promise is followed by a miniature showing the magician in an elaborate building with an open cupola above. He holds a small figure, resembling that of the planetary spirit (without the peacock), and is addressing a group of four winged angels. Clearly, his power has become greater. The information on rings of

Figure 4.2. Rings of Mercury. Vatican, Biblioteca Apostolica, Cod. Reg. Lat. 1283a, f36r.

Mercury continues on folio 36r, with still further triumph depicted in the first miniature. Here the magician, surrounded by the necessary equipment, has actually summoned the spirit into his enclosed room and is directly in conversation, perhaps aided by the special characters shown on the white sheet behind him. The text promises power over the spirit and the ability to give instructions to it. By the end, the magician is white-haired, although he had brown or grey hair in earlier paintings.

This account of what remains of the book of astral magic has shown that it fulfilled two functions. It is both a beautiful picture book and a reference work for professionals. In addition, it was designed as part of a set of such works, giving a comprehensive account of astral magic and its ramifications. It is hard to believe that this research was not intended to be put into practice. As noted above, a full set of magical stones, and their astrological affiliations, is provided in the *Lapidary* – filling the gap in the information given in the *Astromagia*. The illuminations in the *Lapidary* also depict the angels linked to each degree of each zodiac sign, as well as the positions of the main stars in each zodiac constellation. These are important, since they can be cross-referenced to images of individual stars within the discussions of the stones for each zodiac sign. Similarly, the surviving summary of the *Book of Forms and Images* (in Madrid, Escorial, Real Biblioteca, MS h I 16) promises coverage of both celestial images and stones. An important point is that *Picatrix* stands out as both the most theoretical book of all and the most ambitious in the complexity of its information. Presumably this was reserved for a fully trained practitioner of astral magic. A close analysis of Picatrix's instructions for magical rituals linked to Mercury will back this up.

The first relevant section is in Chapter 10 of Book Two. Paragraphs 31–34 (in the translated text) move on to discuss images for Mercury, and we learn that Picatrix instructs that such an image should be of a winged man, accompanied by a cock and with a cock's crest on his head, holding a dart and a shell. However, 'others' have prescribed an image of a crowned noble riding a peacock while holding writing materials – something which is, in fact, closer to the images in the *Astromagia*. Mercury's metals are quicksilver and magnetite. A ring for Mercury should be made on Mercury's day (Wednesday), in the hour of Mercury and with the Moon in Virgo. It should be of magnetite and engraved with the image of an enthroned man with a set of books and accompanied by supporters and students. When used correctly, this ring will confer intellectual and philosophical strengths, as well as influence over people and affinity with bodies of water.

Paragraphs 68 to 72 go on to outline figures or images of Mercury – to be carved into emeralds. Stones of impressive size would be required, as the images are complex and some are accompanied by characters or letters. These are simply to be used when required; but one, described in section 73, is to be made in marble and used as a seal. Wax imprints of this marble seal can then be used to cure a fevered person. All these images are to be made in the hour of Mercury, and most require Mercury itself to be in a specified position. The artist would have to work quickly, yet very precisely; a visual image to copy, like those in the surviving Alfonsine manuscripts, would obviously be useful.

Further practical instruction relevant for magical working in relation to Mercury is given in Book Three of *Picatrix*. Section 8 of Chapter 1 sets out the branches of magic governed by the planet, and adds that all stones that have been worked and engraved can be considered as stones for Mercury. Also useful are the lists of appropriate animals, plants, spices, incenses and colours, which help in understanding the details of the miniatures discussed above. Chapter 5 moves on to magical practice, with the reminder that the magician must first grasp all the relevant information about the planet, its powers, figures and images. Preparation must include bringing together the right image, in the right colour, with the correct suffumigations and offerings. The magician should be wearing clothes in the correct colours and should have practised eating the requisite foods. A suitable location or isolated building is also required, which should be equipped not only with all of the above, but also with an incense burner and cross (the latter two foot high) made at the astrologically appropriate time and of the correct metal (by this stage the magician is expected to know what these would be for any given occasion). The room or building should have an opening to the sky, so that the smoke (after being inhaled by the magician) can ascend to the sphere of the stars at an hour which has been carefully calculated. At this introductory stage it is noted that the planetary spirit will not be perceptible to human senses, although its power, together with its rays, will be drawn down if all is done as instructed. This is comparable to the illuminations of the early stages of the Mercurial magic in the *Astromagia*.

Also related to those illuminations is information given in Book Three, Chapter 7. Here the magician is told to dress like a scribe or lawyer and to wear a special ring of 'solidified mercury'. The Moon should be in conjunction with Mercury, and the magician should be facing Mercury while having writing materials to hand. The incense burner should also be of solidified mercury, and appropriate suffumigations should have been made from nutmeg, holm oak, cumin, cloves, myrtle, almond shells, acacia, tamarisk, grapevines and squinancywort, ground, mixed with wine, formed into tablets and dried. An invocation for Mercury is given, deploying names drawn from several languages and traditions, and also conjuring Mercury through calling upon the angel Arquyl. The magician should display humility by reciting this while prostrate, and should also behead a cock, burn it and eat its liver.

For still further precision, Chapter 9 instructs in the invocation of planetary spirits. The predominant spirit for Mercury is 'Merhuyez' and there are seven lesser and more specific spirits, each of which is named. The rituals outlined here should be performed on Mercury's day and with the Sun in Capricorn, and the magician should have a golden chair and table ready in a special house. Seven goats are also required; and the suffumigation mixes aloes, incense, myrrh, hemlock and elecampane. Inhalation of this would probably have an effect on the state of mind of the magician and may have made the subsequent procedures appear more effective. The goats are to be struck, as eight exotic words or names are called out; they are then to be killed, butchered and suffumigated, before being cooked. This goes much further than the images in the *Astromagia*, but is comparable, in that goats are similarly sacrificed during rituals for Mercury (eleven of the twelve depictions of sacrificed creatures in Vatican, Cod. Reg. Lat. 1283a show goats).

Additional information, also relevant to the texts and images in the *Astromagia*, is found in Book Four of *Picatrix*. Among the guidance assembled here are groups of special characters to be inscribed in preparation for conducting particular rituals. Section 24 of Chapter 2 specifies that the magician should fast for seven days before carrying out the magic, as it was usual. Here the fast should begin on a Thursday, so that the magic could be carried out on Mercury's day. The sacrifice is not a goat, but a black and white cock, recalling the images of Mercury described in Book Two. The angel of Mercury is here named 'Arquil', as in Book Three, Chapter 7; and suffumigation is again required. A set of nine characters is given. At this point no details of the suffumigations are provided, although the preceding section of the chapter has noted that the Indians make use of extremely powerful suffumigations for each planet, called *calcitarat*. Helpfully, Chapter 6 gives recipes for appropriate Indian suffumigations, although the term *calcitarat* is not used. It is here promised that these will enable the magician to attract the planetary spirits and draw them into objects, as well as places. For Mercury, the substances required are henbane flowers, indigo leaves, hazelwort, amber, toad testicles, ammonia, brains of crow, hoopoe and tortoise, and blood from a donkey. These should be mixed, ground and dried, while the magician calls upon the power of the planet; the resulting mass is then shaped into tablets. It is not hard to conclude that carrying out this process and then inhaling the resulting fumes would produce an effect upon the magician, especially when combined with the other parts of the rituals.

The care given to visual images in Alfonso's manuscripts is important in light of the fact that only one of the nineteen known manuscript copies of the Latin *Picatrix* has any illustrations at all.[28] It would seem that later readers of the text either did not require the services of artists or could not afford them. This serves to emphasise the level of Alfonso's interest in the study and practice of astral magic. Of course, it is also a reminder that few other European sovereigns invested quite so openly in exotic and innovative forms of magic. Nevertheless, Alfonso was certainly not the only medieval ruler to be interested in the mysterious celestial figures, images and spirits described and depicted in his books. Equally, he was far from the only king to employ the services of professional experts in astrology and its magical applications. The next chapter will follow the career of another exotic and beautifully illustrated text of 'ancient' astral magic – one that first appears in a version associated with the court of Emperor Frederick II. Even more famous, and more successful, in his own time than Alfonso X, Frederick was called by Matthew Paris *stupor mundi* (the wonder of the world). His desire to master all important branches of knowledge was widely known, and it seems that astral magic was part of this.

5

A GIFT FIT FOR A KING

The Visionary Astrology of Georgius
Zotorus Zaparus Fendulus

In this chapter, we focus on an astrological text that first appeared in a thirteenth-century manuscript, probably produced in southern Italy. In fact, the manuscript shows characteristics that link it to luxury books from the court of Emperor Frederick II (r. 1220–50). Frederick's realms, like those of Alfonso X, enjoyed access to and knowledge of far-flung astral magic. Indeed, both Frederick's territories and his ambitions were even grander than those of Alfonso. Frederick was king of Sicily in 1198, became emperor in 1220 and made himself king of Jerusalem in 1225. Fittingly for a ruler of so many territories, he was celebrated for his wide-ranging knowledge and eclectic interests. One of the most famous scholars attached to his court was Michael Scot (d. c.1236), whose knowledge of, and claims for, magical practices contributed greatly to the legends that grew up around both men. Scot argued strongly for the use of astrological knowledge to make helpful predictions of coming events. His *Liber particularis*, written for Frederick, shows expert knowledge of astral magic and spirit-summoning, together with other branches of magic.[1]

The success of the 'Fendulus' text is shown by the fact that it mostly survives in a group of tantalising and striking manuscripts, produced in France in the later middle ages. All these manuscripts possess extensive sets of extraordinary, full-page illuminations, offering a powerful view of the heavens. These images command attention, even while the text manages to be both authoritative and elusive. Characteristically, it has no clear title, although its prologue describes it as an 'introduction to astrology'. Equally, its 'author' provides an impressive autobiography, calling himself a priest, philosopher and courtier, and recounting his extensive travels and studies of the stars – and yet is quite likely to be fictional. In many ways this authorial figure can be compared to Picatrix, discussed in the previous chapter. Equally, the manuscripts of his book suggest that it, too, enjoyed an enthusiastic reception by elite patrons – and over a very long period. The text itself is brief, but both it and its illustrations focus on the figures and images perceived in the outer celestial spheres by ancient sages, very similar to those in the books of Alfonso X. In the case of the Fendulus manuscripts, however, the emphasis is on the images: there is far less textual instruction on how they might be used in practice.

In a way, the elusive identity of the author/editor does not matter, since the text is stated to be a translation of the 'book of Abumazar' (Abū Ma'Shar) rather than a new creation. That attribution would carry weight for any medieval reader interested in astronomy and astrology. 'Abumazar' (Albumasar) was acclaimed as the author of two fundamental textbooks on the subject, which had been translated into Latin in the twelfth century.[2] However, the details given about the translator's identity and credentials cast doubt upon the truthfulness of the whole account. The author/translator gives his name as Georgius Zotorus Zaparus Fendulus, and a 'portrait' is given in the oldest known manuscript of the work, which was probably produced in southern Italy towards the middle of the thirteenth century. Here he appears as a grand and somewhat exotic figure, dressed in impressive robes. This manuscript is now in Paris, Bibliothèque nationale de France, as Lat. 7330, and the full-page image of Fendulus prefaces his text.[3]

To be fair to Fendulus, at least one individual is known to have combined the roles of philosopher and courtier in a southern Italian court in the relevant period. This was Theodore of Antioch, recorded

Figure 5.1 Georgius Zotorus Zaparus Fendulus.
Paris, Bibliothèque nationale de France,
MS Lat. 7330, f1r.

as 'imperial philosopher' to Emperor Frederick II from 1227 to 1236, and described by chroniclers as successor to the reputed magician, Michael Scot.[4] Theodore's name links him to the eastern Mediterranean, and suggests a scholar with expertise in Greek and Arabic. He would thus be able to provide for Frederick, whose curiosity about the natural world and the heavens is well recorded, a version of the knowledge made available to Alfonso X of Castile and Leon. A later source claimed that Theodore visited the court of the 'sultan of Babylon' and was instrumental in the acquisition of a copy of an important astrological compendium, *The Book of Nine Judges*.[5] Frederick ruled the multicultural territory of Sicily, which had trading and crusading links to the Crusader states, including the Principality of Antioch.[6] If this was indeed the court for which Fendulus' work was intended, then the emperor's possessions and patronage in northern Europe, as well as his fame, will have helped in the spread of copies. Similarly, the status of a text offering esoteric knowledge from the imperial court may have encouraged patrons to invest in copies, even though the number of complex images would make them very expensive.[7]

The autobiographical prologue goes on to narrate Fendulus' arrival in 'Cethyn' after frustrated attempts to observe the stars in his own (unstated but cloudy) location. Better visibility convinced him that his copies of earlier Greek and Latin works were faulty, and that further travels were needed to gain access to Babylonian records of the stars. His further research apparently included translating astronomical and astrological materials from both Persian and 'Chaldean', and brought him more accurate knowledge of 1,029 stars. The suspicious might note that this is very close to the total of 1,028 listed in the updated tables produced for Alfonso X.[8] He omits the usual information concerning the constellations recognised by classical astronomers. Despite this, he claims to have depicted, and to present in his volume, together with his translation of the 'Book of Abumazar', all these stars. Many of the details of this account are dubious, beginning with the statement that it was necessary to translate the (unnamed) astrological text from Persian: the original would have been in Arabic, as that was the language used by Abū Ma'Shar. Arabic was also the dominant language in the places Fendulus claims to have visited, thus translation from Persian would not have been required. The cities named as Cethyn and Carabolis are equally puzzling, although Babylon and Damascus were names well known in medieval Europe. All this goes rather against Fendulus' claims concerning the care and accuracy of his work, although it is certainly true that he has presented his findings in a concise text which is very fully illustrated.

Recent research has identified the text offered by Fendulus as consisting mainly of selective extracts from the *Great Introduction to the Science of the Judgements of the Stars* by Abū Maʿshar, in the Latin version of Herman of Carinthia, dated 1140.[9] This rather free translation was less popular than the more literal version produced by John of Seville in 1133. Both were very much longer than the text presented by Fendulus, which gives information only on the natures and powers of the signs of the zodiac – concentrating on their little-known subdivisions, the 'decans' – and on the changing powers of the planets as they moved through those signs. No guidance is given as to how to discover the positions of the planets on any given day, suggesting that this work is aimed at patrons who wish to gain privileged insights into the 'secrets of the stars', rather than practising astrologers.

Introductory information is first given as to the elements (fire, air, water and earth) and qualities (warm, cold, moist, dry) which link material life on Earth to the planets, stars and signs. This begins on folio 2r of Lat. 7330 with the impressive assertion that the ancient experts, Ptolemy foremost among them, had established the dimensions, qualities and movements of the celestial sphere. Those who have questions or wish to know more are referred to Ptolemy's *Almagest*, here given its Greek title – spelt out in a version of Greek characters – as well as the Arabic title by which it is generally known. The translator's credentials are thus emphasised. Fendulus is here echoing Abū Maʿshar, who also referred to Ptolemy as a fundamental source – although the exact number of stars given varies slightly.[10] As promised, the details which follow are also based on an early section of the *Great Introduction*.[11] This part of the Fendulus text concludes by pointing out that there are four groups of three signs making up the zodiac, each group being linked to one of the elements. First place is given to the signs associated with the life-giving element of fire – which are Aries, Leo and Sagittarius. The water signs – Cancer, Scorpio and Pisces – come last. In accordance with Abū MaʿShar's philosophical explanations of the innate connections between celestial bodies and earthly phenomena, the explanation of this order expounds on the powers of the elements in relation to earthly life. This bond is central to the argument that the stars, signs and planets can be linked to events and creatures on Earth.

More specific guidance on each sign in turn is drawn from a later section of the *Great Introduction*.[12] The signs are discussed and illustrated in their established order, beginning with Aries, the sign linked to the spring equinox, and occupying the first 30° of the circle of the ecliptic. The very fact that the text was drawn from a twelfth-century translation of a book that was already established as a classic in its field shows that none of the information provided by Fendulus was actually a new discovery. Experts in the rapidly spreading, more technical, Islamicate astrology will already have been familiar with the idea of the decans as 10° divisions of the zodiac signs.[13] The decans also played a role in the expanding field of astral magic, and are expounded, for instance, in the wide-ranging treatise *Picatrix*.[14] Book Two of that work set out the 'celestial images' of the constellations as delineated by Greek astronomers and then the star-groupings assigned to the decans of each zodiac sign. Like Fendulus, the compiler of *Picatrix* makes use of the account given by Abū MaʿShar.[15] Moreover, the idea that the planets (including the Sun and the Moon) have increased power in certain zodiac signs (known as their houses and rulerships) and diminished power in others (their detriments and 'falls') was expounded by Ptolemy and other authorities, and thus was even more widely known.[16] As noted above, the readership addressed by Fendulus' work clearly was not assumed to consist of expert astrologers or astronomers. Nevertheless, credit should be given for the attempt to provide original and detailed images for all of the three sets of 'figures' perceived in each decan of each sign.[17]

A further indication of the expected readership is that no detailed, overall list of the names and locations of Fendulus' newly ratified stars is provided. A broad account is given in the introductory section of the identification and labelling of constellations by the Indians, Persians, Chaldeans and Arabs, with figures for the numbers of stars involved; but again, this sets out established knowledge. Perhaps even more surprising is that the illustrations of the work, fascinating and detailed as they are, make no attempt to indicate the locations – or even the presence – of stars within the various views of the heavens depicted. The reader is left with a strong impression of the complexity of the heavens on the one hand, and of the powerful connections between human affairs and celestial bodies on the other, but with no practical way of making use of the information. Owners and readers of these manuscripts would thus have access to a rich and strange gallery of images linked to stars and planets, combining details from exotic myths and legends with fragments of everyday life. They would presumably also be able to consult practising astrologers for interpretations of the images, as well as 'judgements' offering individually tailored guidance as to how these celestial entities and forces would impact on their own activities. To demonstrate the experience of encountering these images, and the rather sparse texts accompanying them, it is now time to look at some examples.

At the beginning, the sign of Aries is depicted on folio 6r of the oldest manuscript in its 'classical' guise of a ram, here hovering over a field of red, blue and white (spring?) flowers. The name Aries is written in gold letters at the top of the picture's frame, and small, white stars occupy the blue sky behind the ram; but there is no suggestion that the symbolic image itself is made up of stars, or depicts their relative positions. This picture is thus different from many of the Carolingian works discussed in Chapter 1 above, despite its classical style. It is theoretically possible that this absence is related to the growing acceptance by astronomers and astrologers that there were two zodiacs: the first was based on the constellations after which the signs were named, while the second was based on 30° sections of the ecliptic. This separation of the signs and the stars was made necessary by the fact, noted by astronomers at least from Ptolemy onwards, that the positions of individual stars and star-groupings, when given as longitudes on the ecliptic, changed over time. Left uncorrected, this would mean that the Sun's entry into the sign of Aries and the day of the spring equinox would not coincide. The solution was to separate the zodiac based on the stars from the zodiac based on fixed sections of the ecliptic. This in turn necessitated the acceptance of the cosmos having nine spheres, going outwards from that of the Moon through those of each of the other planets, and then the sphere of the fixed stars, before terminating in the ninth or *primum mobile*. The last-named contained no planets or stars, but had the function of imparting movement to the lower spheres. It would follow that the section of the outer zodiac occupied by Aries would still have the same nature as the star-based, inner Aries – but would contain no actual stars. Thus, an image simply of the symbol for Aries would be appropriate. Moreover, the version of Aries beyond the stars, since it was closer to the outermost *primum mobile*, would potentially be more powerful than the 'lower' Aries.

A brief account of the problem and solution was given for thirteenth-century students in the Latin world in Sacrobosco's textbook on astronomy, *De sphera* (*On the Sphere*).[18] The problem and its consequences are also outlined in Abū MaʿShar's *Great Introduction* – just before the details concerning signs and their divisions, which constitute a major part of the Fendulus text. Anyone familiar with the work of Abū MaʿShar will thus have known about it. The great astrologer noted: 'We find the stars have moved several degrees between the days of Ptolemy and our own days, and because the stars have left their positions, the figures [of the constellations] have strayed . . .'[19] This phenomenon was important because, as the *Great Introduction* goes on to explain, the star-groupings associated with specific sections of each sign have moved in relation to those sections. In fact, Abū MaʿShar had recalculated the relationship between sign-sections and star-groupings, and made them accurate for his own time (849–50).[20] However, the Fendulus text does not expound such subtleties: its text is kept to essentials for each zodiac sign and each section of each sign.

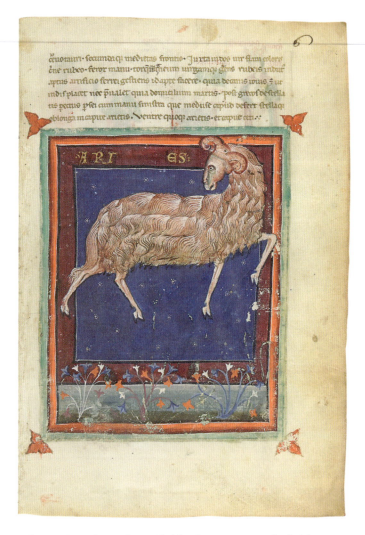

Figure 5.2. Aries. Paris, Bibliothèque nationale de France, MS Lat. 7330, f6r.

The details concerning the nature of the sign Aries, accordingly, are both brief and, as would be expected, taken from the *Great Introduction.* They occupy most of folio 5v and the top six lines of folio 6r. The sign of Aries, the reader learns, is fiery in nature (linked to the element of fire and part of the Fire 'triplicity'); bitter in taste; straight or upright in appearance; bicoloured; and dual in form. It also brings increasing length of daylight and rising times of less than 30°.[21] The next statement is more surprising. This is that:

In its first decan, according to the Persians, there rises a woman whose name is Daughter of Splendour. After her come the tail of a sea fish and the first part of the Triangle. Then the head of the stag-bull (formed from a stag and a bull); and then a dog-headed figure holding a torch in its left hand and a key in its right.

As promised by Fendulus, images corresponding to these forms are given, grouped together in the upper register of the full-page illumination on folio 6v. As the illustration below shows, each image is accompanied by a rubric in red, helping the viewer to identify it and link it to the text. However, there are also indications that the designer of the illumination was challenged by the task. This is likely to have been caused by the brevity of the text supplied, which omits some important information given by Abū MaʿShar about the decans. This was that ancient authorities had divided up the signs in different ways and with differing terminology. Decans were one such division, recognised by the Persians, the Babylonians (or Chaldeans), the Egyptians and the Greeks.[22] Abū MaʿShar also explained, in rather oblique terminology, that the forms assigned to decans by the Indians, Persians and Egyptians did not depend on the permanent presence within those decans of specific stars. Instead, they are based in some systems on the meanings attached to the stars and star-groupings, while in others they can be entirely

Figure 5.3. The first decan of Aries. Paris,
Bibliothèque nationale de France, MS Lat. 7330, f6v.

symbolic. In other words, the names and forms do not necessarily
relate to actual stars. Herman's Latin translation offers the further
note that this means that the same images or forms are always found
in the same decans, regardless of the positions of the stars, and that
they have the status of being '*intersignia*' linked to the higher sphere
(the *primum mobile*) itself. If accepted, that would make these forms

156

more powerful than the other astrological bodies, rendering this information potentially important.

More detail on the differing views of decans is found in a later section of the work.[23] The three decans assigned to each zodiac sign are expounded in turn, with the star-groupings and/or forms perceived in each by the Persians, the Indians and the Greeks. Thus, the illumination on folio 6v represents the parallel sets of figures perceived in the first decan of Aries by the three sets of astronomer-astrologers. The images themselves certainly communicate an unusual, and little-known, view of the heavens. However, their relationship to the text is made somewhat awkward for the reader, due to the confusion of the illustrator mentioned above. It would be logical to assume that each horizontal section of the illumination will depict one set of descriptions, with the Persians coming first; but this is not quite what happens. For it is clear that the list of Persian forms does not include the standing figure, dressed in red and white, at the far right of the top register. Reference back to the text on folio 5v finds that the Indians assigned the whole of this decan to a dark-complexioned man with red eyes, a large body and fierce spirit, standing upright and wearing a white linen cloth. This is presumably the concept depicted here, although the correspondence is not complete. By contrast, all of the middle register is given over to forms discerned by the Greeks, which are based on the constellations as described by Ptolemy. This list is long and detailed, and includes portions of recognised constellations, treated as parts of actual bodies, such as parts of the constellation of Cassiopeia (the buttocks, knees and left hand), parts of Andromeda (the lower back, buttocks, thighs and lowest part) and the intestines of Cetus (a large constellation usually depicted as a sea monster).

The fact that the rubricator and illuminator attempting to illustrate this text were somewhat challenged by it is not only shown by their lack of distinction between the Persian and Indian accounts of the decan. It can also be seen in their failure to illustrate the

Figure 5.4. The second decan of Aries. Paris,
Bibliothèque nationale de France, MS Lat. 7330, f7r.

intestines of Cetus. Instead, the lowest register of the illumination moves on to depict the first part of the 'Persian' set of figures to be found in the second decan of Aries. There are three of these. The text names them as the middle of a sea fish; the middle of the Triangle; and the middle of the stag-bull. As the text goes on to expound, these are to be followed in the same decan by figures including the 'sword' or 'scimitar' of Perseus. The latter is given a long description, whose terminology is confused in places. This has led the artists to show, in the middle register of folio 7r, a stag (*cervus*) and a harp (labelled 'harp of Perseus') which should not be there. The figure on the right, a female centaur, does illustrate the figure described by Indian astronomers. A final detail is that the portions of 'Greek' constellations usually shown as human figures, occupying the lowest register on folio 7r, are here depicted as recently severed body parts from which blood is issuing. This adds considerably to the drama of the image.

Overall, what this analysis has shown is that the contents of Fendulus' treatise posed problems for the illuminators, who worked hard to provide visual equivalents for the individual figures described – but privileged fitting them neatly into a three-tier pictorial layout over complete faithfulness to the text. The impression given is that they were not familiar with the astronomical and astrological phenomena that they were required to depict.

That impression makes it necessary to check whether this lack of clarity continued throughout the volume. For that reason, the sign of Libra will also be examined. The text for Libra is found on folio 20v and states briefly that the sign is temperate, dual in colour and form, autumnal and upright. Details concerning hours of daylight and rising times are not given. The Persian figures for the first decan are listed as 'an angry man carrying scales in his left hand and a lamb in his right; books with a third part of the knowledge of those whose name[s are] Karaz *mune* [the text here is not entirely clear]; a musician on a horse, striking a drum and playing a reed pipe; the head of the dragon; the first part of what the Greeks call the Great Bear; and a ship's keel [*carina*]'. At this point the scribe, perhaps confused by the reference to Greek astronomy, went straight on to the figures discerned by the Greeks. The Indian description, given in the *Great Introduction*, of a market trader with scales is omitted. The Greek figures are listed as the middle of the tail of the Great

Figure 5.5. The first decan of Libra. Paris,
Bibliothèque nationale de France, MS Lat. 7330, f21v.

Bear (Ursa Major); the middle of the Virgin with an ear of corn in
her left hand; the tail of a stag (*cervus* – an error for *corvus* or crow);
part of the tail of the southern snake or Asura; the right shoulder of
Centaurus; and part of Pegasus.

The illustration on f21v (Fig 5.5) shows what the illuminators
made of all this. It is clear that they did their best to follow the text,
although they, too, found it obscure in places. The top register

depicts the angry man (labelled as such) with his accoutrements; a three-part book (?) labelled 'Karaz mune'; a musician riding a horse and playing (oddly) a pipe labelled as golden; a separate drum; a dragon's head; a small open boat (labelled *carina*); and an enigmatic red square labelled as 'part of Ursa Major'. The last of these may suggest that the illuminator was unfamiliar with the concept or constellation of the Great Bear; yet the first figure in the middle register shows the clawed back leg of a large, shaggy creature and is labelled 'tail of Ursa Major'. Like the text, the illumination omits the 'Indian' figure. As already noted, the idea of the tail of the Great Bear posed no problems for the artists, although the rubric for this image calls it *coxa* or hindquarters, rather than *cauda* or tail. Similarly the next figure follows the text in depicting the rear leg and tail of a stag; however, the rubric gives both *cauda* and *coxa*. The handling of the portion of the figure of a young woman, or *virgo*, also departs from the detail of the text. Even stranger is the inclusion of a right arm labelled 'right hand of the virgin/young woman' – which does not appear in the text. The artists stayed with the theme of arms in depicting the 'right shoulder of Centaurus', since a whole arm and hand are shown. Confusion was also caused by 'part of the tail of the southern snake or Asura', since two separate images show a snake's tail suspended below part of a pink, winged dragon, with the latter named as '*asua*' [sic]. Pride of place in this middle register goes to an image of Pegasus as a centaur, holding a knobbed stick and a plant. This may be a confusion caused by the text's description of the 'right shoulder of Centaurus' closely followed by 'part of Pegasus'; but nothing explains the objects being held. At this point the figures for the first decan finish and the bottom register of the illumination is devoted to the 'Persian' figures for the second decan, as happened with Aries.

This detailed comparison suggests that the text and images for the decans could be problematic for both scribes and artists, especially as the artists seem to have had to work directly from the text.

The text itself is set out in a regular book hand, without very many abbreviations, but with occasional slips that have mostly been corrected or noted in the margin. The artists and rubricators, however, while attempting to produce images closely based on the text, are sometimes confused by it – and sometimes depart from it for reasons that are not immediately clear. The very fact that the Persian and Greek figures were mostly based on portions of star groups, while the Indian figures were symbolic, causes some lack of clarity, as Abū MaʿShar had pointed out.

The next question is whether the relationship between text and images was simpler in the case of the much more familiar subject of the planets. These are dealt with in a second part of the volume, presented almost as a separate treatise. What happens is that the details of the decans end on folio 35r. A puzzling gap was then left between 35v and 40v, although later hands have added drawings and tracings on folios 36r, 36v and 37v. These originally blank folios suggest that some part of the plan for the volume was not completed. On folios 40v to 41r there is a short text on the origins of varying views of the planets and their powers, largely – though not entirely – drawn from the *Great Introduction*. This is followed on 41v by a full-page painting of an author, presumably 'Albumasar', which appears to be by the same hand as the image of Fendulus. This part of the volume is clearly designed to accompany the materials on the decans, and appears to be the work of the same team of scribes and illuminators. Albumasar's authorship of this section is emphasised by the fact that the details of Saturn, the outermost and first planet to be dealt with, are on folio 42r, facing the 'author portrait'. The approach to the material is similar, emphasising that the planets have affiliations to the same elements and qualities as the zodiac signs.

It was, of course, an established doctrine of both astronomical and astrological teaching that the planets circled around the Earth, each following an individual path through the zodiac signs, and moving at a distinctive speed. The Sun and the Moon were treated as planets for this purpose. The planets were also traditionally identified with individual gods and goddesses of the classical pantheon, although they were no longer recognised as deities in the medieval period. Indeed, the Fendulus text and images emphasise their links with human life and human activities and are unusual in suggesting that they experience changes of fortune similar to those occurring in human lives.

Information on Jupiter, the second planet in the traditional order, begins on folio 44v. It is extracted from two different portions of the *Great Introduction*, as is the case for all the planets. First come the statements that Jupiter is hot, moist, temperate and agreeable in nature. Following on, we learn that he has natural power to support life, both physical and spiritual, and that he is linked to progeny and to great individuals and prelates. He rules the left side of the body, especially the ear and (perhaps) the liver. He is dignified and noble in form, wise and penetrating, and powerful in understanding and interpreting things seen. In terms of human life, he is especially linked to laws, religious observances, honesty, resilience, good judgement, peace-making, faithfulness, humility and obedience. When negatively affected, he can indicate acting without advice, hastiness in business affairs and recklessness. Among human beings, he is most connected to those in powerful positions, and brings success and victory, together with benevolence, good judgement and patience. Less positively, he also signifies boundless ambition and a burning desire for riches. However, this wealth is likely to be used well, and especially to support amity, positive contacts and good behaviour on a social level.

The list of Jupiter's attributes and powers is followed on folio 45v by a full-page illumination of the planet/deity as a thirteenth-century ruler or magnate, seated on a richly cushioned seat of Byzantine appearance and making a gesture signifying the act of judging or making an announcement (Figure 5.6a). There is little to

jupit́ mafcul᷎ natura caud᷎ humid᷎. dulcíf t᷎par᷎ equuf. ꝗabtt du
as dmuf fagittariu díe. pisem nature.

pifcts

Figure 5.6a. Jupiter with Pisces and Sagittarius.
Paris, Bibliothèque nationale de France, MS Lat. 7330, f45v.

inform the uninitiated that this is an image of a planet. Perhaps equally puzzling are the depictions of a centaur and two fishes flanking the throne of Jupiter. The fish are joined by a thread, as was customary in depictions of the zodiac sign of Pisces, and are labelled 'Pisces'. The centaur has no accompanying label, but is clearly important, since so much space has been allowed for his depiction that Jupiter himself is placed off-centre in the composition. Brief but helpful information is provided by the text above the illumination, which states that Jupiter is masculine, warm, moist and temperate; also that he has two 'houses', which are Sagittarius by day and Pisces by night. The difference in scale between the large figure of Jupiter and the small ones of the zodiac symbols works against the astronomical and astrological concept that Jupiter appears in the sky as a small body transiting (i.e. passing through) the 30° sections of the ecliptic occupied by the zodiac signs, and that his power fluctuates significantly as his position changes.

However, that message is perhaps strengthened by the fact that three further full-page illuminations follow, each depicting Jupiter accompanied by relevant zodiac symbols, and each with a brief text giving extracts from the relevant information (though not necessarily all of it). These show Jupiter in dramatically contrasting states. When under the influence of Virgo and Gemini, he stumbles and begins to fall, still wearing his crown, but clutching a broken sceptre and no longer enthroned. In the next image, his dignity is restored by the influence of Cancer, his Exaltation, and his throne is perhaps even grander than that shown on folio 45v. However, things end on a sombre note, with Jupiter in headlong Fall in Capricorn. Here his attributes of power are lost, his eyes are closed, and he appears almost about to be trampled by the symbol of Capricorn – here simply a goat, rather than the usual hybrid creature (Figure 5.6b).

The approach is similar for each planet, following the order of the planets' perceived distance from Earth, beginning with Saturn and proceeding through Jupiter, Mars, the Sun, Venus and Mercury,

Indi acpse dicunt in yma parte sagittarii, remotu asole, xx parnb; que
niente sole inxr gdu leonis haber pactu getagnmoth suu. Sua affinita
tes 7 montes. haber incapcorno inxij gdib; sic aincte res piclitantur
ut uidesi.

Figure 5.6b. Jupiter and Capricorn.
Paris, Bibliothèque nationale de France, MS Lat. 7330, f46r.

Figure 5.7. The chariot of the Sun. Paris, Bibliothèque nationale
de France, MS Lat. 7330, f51r.

before concluding with the Moon. A classicising touch is that the
Sun is given a full-page illumination showing his 'chariot of fire',
drawn by two horses, and accompanied by a text linking him to
Apollo (Figure 5.7). This emphasises the Sun's importance, since it
means that he has five images, rather than the usual four; but the
concept seems to have confused the team producing the manuscript.

167

Figure 5.8. The Moon. Paris, Bibliothèque
nationale de France, MS Lat. 7330, f69r.

The artists make an attempt to show the chariot of the Sun as a classical two-wheeled chariot – but the Sun himself is missing, there are no flames, the horses of Apollo are down on the ground, and a stray textual extract at the top (concerning the sign of Leo) belongs with the preceding image rather than with this one. Nevertheless, however great the power of the Sun, his sequence still concludes with his downfall and death – in this case in Libra.

The other great luminary, the Moon, has only the usual four illuminations, though she is linked to the Sun in the textual commentary and is similar in having only one 'house' (the watery sign of Cancer). Nevertheless, the text emphasises the significance of the Moon for anyone attempting to read the heavens on any given day or night. The text accompanying the illumination of the Moon in her pomp in Cancer also strikes a classicising note by linking her – rather ominously – to Proserpine and Hecate. This is echoed in the image, where the Moon appears as a combination of a thirteenth-century Byzantine queen and a pagan deity crowned with a crescent moon. She holds a large, floral sceptre in her left hand; in her right is not an orb, but a gold plate or disc containing a blue-and-white symbol, perhaps relating to her power over the oceans (Figure 5.8).

Nothing is known for certain about the original owners of this manuscript, although its styles of script and illumination link it to other de luxe manuscripts from the Kingdom of Sicily. It is significant that it appears unfinished and contains a set of images depicting courtly feasting and celebrations (folios 80v to 85r), following a text commenting on the ups and downs of human fortune. The latter is illustrated by a set of full-page drawings and paintings of the Wheel of Fortune, accompanied by a king, a queen and musicians (folios 72r to 79r). These factors together suggest that the manuscript may have been made as a presentation gift to mark a specific occasion. One combination of events that might appropriately link triumph and disaster occurred in 1235 – the year that saw the marriage of Emperor Frederick II to Princess Isabella of England (daughter of King John and Isabella of Angoulême), as well as Frederick's final triumph over his ambitious and rebellious son, Henry VII, who had caused conflict and war in Frederick's German territories.[24] However, the tone of the volume is not really what might be expected of a wedding present.

Overall, this manuscript inspires both wonder and a certain foreboding. On the one hand, the celestial bodies, normally presented as remote and inhumanly powerful, are here shown as having bodily forms, through which they experience very physical – and almost human – triumphs and disasters. On the other, the heavens are presented as bewilderingly complex, and multi-layered in significance, in a way that emphasises the special skills and knowledge required to interpret them correctly. One thing is clear from the current state of the volume: it was well looked after. It has few stains, marks of use or doodles. It is thus likely that it was kept in the possession of wealthy individuals, who made it available only to selected readers. It also appears to have inspired a select group of later copies, most originating in France or the Low Countries, and involving expensive work by expert scribes and illuminators. The earliest known copy to have survived is now in the British Library, as Sloane MS 3983.[25] Like Lat. 7330 before it, this volume can be dated and located only by the style of its script and illuminations, which suggest a workshop in fourteenth-century France or Flanders. It is so closely related to Lat. 7330 that it provides evidence that the earlier book must have been involved in its production; however, the images depart from those of Lat. 7330 in ways which suggest that Sloane 3983 was not copied directly from it. That in turn suggests the existence of at least one intervening copy.

The Sloane copy shows reduced interest in the mysterious figure of Fendulus, omitting any portrait of him as author. The expenditure on illuminations is more restrained, with the images being just as large and enticing, but carried out in the less expensive medium of tinted drawings. These have been updated and relocated to the world of their patron and creators, and considerable attention is given to details of clothing and architecture. However, the decorated initials are more restrained than those in the older book, giving the text something of the appearance of a scholarly work, even while the images are more 'worldly' and less celestial. This is slightly puzzling, but may suggest that the volume was produced for a patron who could call upon the services of university-trained experts to read and interpret the text, while wanting recognisable versions of the illustrations.

Once again, the artists do not appear familiar with the material they were illustrating. Their updating of the images into a more fashionable style included the omission of the rather 'old-fashioned' frames and backgrounds of the zodiac images in Lat. 7330, leaving the symbolic creatures simply set against a ground of plain vellum. More puzzling is the handling of the images for some of the decans – including the first decan of Aries. The upper register of the illumination depicts the same figures as Lat. 7330 – but the mysterious man in red, draped in a white cloth, who represented the Indian symbol for this decan is here integrated into the illustrations of the Persian figures. His red garment is omitted, leaving him simply draped in a pale cloth, and he appears to be involved in conversation with the cynocephalus standing beside him – something which entirely contradicts the text. His semi-nakedness and pale skin tone depart from both the text and the older painting, and are harder to explain than his integration into the Persian images, since the latter was already suggested in Lat. 7330. Given that the artists of Sloane 3983 appear not to have gone back to the text, this new handling of the image raises different possibilities. One is that the Sloane artists simply arranged the figures more elegantly on the page – something that could lead to the production of the integrated pair of figures in conversation. However, a comparison of the illustrations for the final decan of Aries tends to support the idea of an intervening stage in the handling of the treatise.

As Figures 5.9a, 5.9b and 5.10 show, the later version of this decan again reveals an effort to clarify the layout and to increase the elegance of the figures. This is especially clear in the upper register, where some star-groupings specified by the text and included in the Lat. 7330 illumination have been completely omitted. The text again starts with the figures perceived by the Persians. It lists 'a man called

Figure 5.9a. The final decan of Aries. Paris, Bibliothèque nationale de France, MS Lat. 7330, f7v.

Figure 5.9b. The final decan of Aries. London,
British Library, Sloane MS 3983, f5r.

Fasius; his son [a misunderstanding of the original Latin text]; two
horses; the head and middle of Pisces; the rear of Trigon; the tail of
Cervotaurus, following the middle of Frons'. Once again this has
caused some confusion for the artists of Lat. 7330, who attempt a
literal depiction, but make little sense of the 'rear of Trigon'. It is
perhaps the portion of a red-and-yellow structure placed above the
head of the boy – but that could equally be the 'middle of Frons',
unless the latter has been transformed into the arched doorway
behind the horses, which bisects Pisces. There is no obvious

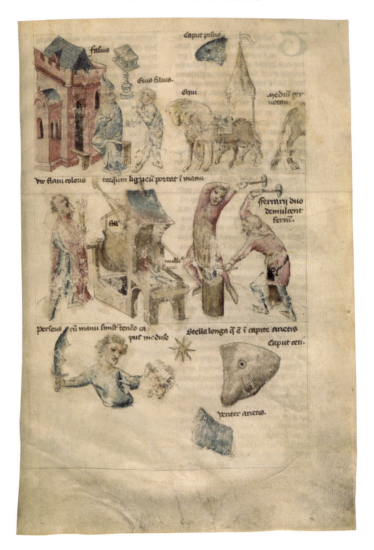

Figure 5.10. The final decan of Aries. New York,
Pierpont Morgan Library, MS M785, f5r.

explanation for the bleeding rear half of a bird that is depicted below
the tail of Cervotaurus. The designer of Sloane 3983 took a bold
approach to this illumination and omitted not only the rear part
of the anonymous bird, but also the rear part of Pisces. The

red-and-yellow structure is transformed into a complex bookstand or press, with no accompanying label; while the structure behind the horses is much closer to being triangular in form, even though this leaves part of Frons unaccounted for.

Once again, the 'Indian' symbol is significantly changed in Sloane, with the very red figure, described in the text as holding a torch and a staff, now depicted as more human in appearance and holding only a torch. Equally puzzling are the changes to the bottom register, where the 'Greek' figures are depicted. The Sloane image of Perseus is closer to the text in showing only the upper part of his body; but the decapitated head is much less recognisable as that of Medusa. The head of Cetus and the belly of Aries have both been transformed, with the head more clearly that of a sea monster and the belly made to accord with the depiction of Aries the ram.

Two things are thus clear. The first is that the illuminations in Sloane are not simply based on those in Lat. 7330; and the second is that the changes are only occasionally the result of going back to the text. Some actually move further away from it. Perhaps most surprising are the images for the Sun and Moon. The handling of the Sun is novel in changing the age of the male figure representing the planet from one image to the next. When in his 'house' of Leo, as shown on folio 39v, the Sun is a young man, holding a disc of swirling red and orange; but in his 'Exaltation' of Aries, on folio 41r, he is a mature man wearing a crown. The image of the chariot is omitted completely, and folio 40r is left blank. The real shock comes with the depictions of the Moon, starting on folio 47v, since the usual female figure is here replaced by that of a young man. This is despite the fact that the adjectives applied to the Moon in the Latin text remain feminine. Indeed, the young man depicted in his 'house' of Cancer has no regalia or insignia relating to the Moon, apart from a small line drawing of the Moon's face engraved on the leather bag he is wearing. Subsequent images of the Moon's decline and Exaltation show an older figure – but still a man. The absence of the

Sun's chariot and the transformation of the Moon from queen/ goddess to secular young man have a double effect: at one level, they reduce the classicising elements in the images; on another, they strengthen the tendency to depict the planets in terms of contemporary human life.

Taken together, this evidence strongly suggests that at least one handling of Fendulus' treatise (and probably more) intervened between the thirteenth-century Italian original and the fourteenth-century northern French version. This is important since it means that illustrated copies of this mysterious work were made for several patrons, and that attempts were undertaken to reinterpret the images for a new milieu. The updated copy suggests a more direct relationship between human life and the ever-changing states of the planets. That in turn makes it less surprising that the work enjoyed a period of prominence in the French court during the fifteenth century, at a time of military and political problems for France and of significant debate about the legitimacy of astrology. It is certainly the case that devastating reversals of fortune and unexpected twists of fate occurred rather frequently at this point.

The French King Charles V, who had brought stability and success in the ongoing war with the English, following the capture and imprisonment of his father, John II, had died in 1380.[26] His son, Charles VI, was then aged only eleven. That gave the young king's ambitious uncles the opportunity to become regents, a situation that recurred when Charles VI suffered a severe breakdown in 1393. Worse was to come, when a civil war broke out between the duke of Orléans, brother of Charles VI, and the duke of Burgundy. Following the disastrous defeat of the Battle of Agincourt in 1415, this tension increased, and a Burgundian coup led to the occupation of Paris and the flight of the dauphin (the future Charles VII) in 1418. The assassination of Duke John the Fearless of Burgundy in 1419 was followed by a successful alliance between the Burgundians and the English, the marriage of Henry V of England to Catherine, daughter of Charles VI, and the disinheriting of the dauphin in 1421. However, further turns of Fortune's wheel saw the death of both Henry V and Charles VI in 1422, the miraculous military successes of Joan of Arc and finally the reconciliation of Charles VII and Duke Philip the Good of Burgundy (in 1435). This long period of war and turmoil took a heavy toll on the people of France, reflected in rebellions and uprisings. Guidance concerning the future would thus be even more important than usual, and the rather ominous imagery of the decans and the overthrown planets would perhaps suit the political mood.

The idea that Fendulus' work was in demand at this difficult time is based on the fact that two of the surviving copies are connected to the French court. Both also have links to the Sloane manuscript, as well as new features of their own, suggesting that patrons were sufficiently interested in the work to want special details, rather than a simple copy. The older of the two is now in New York, as Pierpont Morgan Library, MS M785.[27] In 1403, it was presented to the famous bibliophile, Jean, duc de Berry, brother of King Charles V of France and a regent during the minority of his nephew, Charles VI. The donor of this impressive gift was Lubertus Hautschild, abbot of St Bartholomew's, Eeckhout, and counsellor to both Jean de Berry and his brother, Philip the Bold of Burgundy. It is likely that Lubertus would be able to expound the concepts so dramatically presented in the images, since he was known for his expertise in chronology, astronomy and astrology. His status as a counsellor and ambassador in both France and Flanders seems to have been related to the success of his astrological prognostications. He was also known for his ability to construct teaching aids for these complex subjects, such as a model which demonstrated the movements of the planets and stars. As might be expected, he had an impressive book collection on these topics, which presumably included a copy of the Fendulus text, available to be used as a model. An inscription states that Lubertus himself 'ordered' the figures and images in the copy of Fendulus' work to be presented to Jean de Berry.[28]

This volume is evidence of the significance accorded to Fendulus' mysterious images both as insights into the nature and effects of the stars and planets, and as material for an impressive work of art. Despite the difference in date, the book is strikingly similar to the Sloane manuscript. Even the layout and decoration of the text, from the prologue onwards, are closely related. In the case of the images of the signs and their decans, the relationship is, if anything, even closer – including the medium in which the illuminations are executed – even though the artists are clearly different. Once again, the symbols of the zodiac signs are set against a plain ground and have no frames, while the pose of the ram of Aries is identical. A comparison of the figures depicted in the third illumination of the decans of Aries, on folio 5r, is still more convincing. Once again, details of clothing, objects and architecture have been updated to accord with fashion, here that of c.1400. But the figures and captions themselves are clearly based on those in Sloane 3983 and depart from the text in the same ways. The only difference in content is that the 'Indian' figure of the man in red is now given a garment in that colour. This is a correction likely to be based simply on the caption, which calls him 'flame-coloured'.

The closeness to the Sloane volume extends to the depictions of the Sun and Moon, which are very similar, including the transformation of the Moon into a male figure. The designers of the de Berry book followed Sloane in having no image of the Sun's chariot; but they did not leave a blank folio where that picture might have been. Thus, the Sun has a simple set of four images, like the other planets. This once again supports the idea that the duke's book was closely based on Sloane, since the duke himself had no objection to the concept of the Sun as a classical deity driving a chariot. The evidence for that is provided by the famous illuminations accompanying the calendar in the duc de Berry's *Très Riches Heures*. At the top of each calendar image is a depiction of the Sun as a human figure, wearing a multi-tiered crown and driving a four-wheeled chariot drawn by two winged horses.[29]

The copy from the later fifteenth century, now Paris, Bibliothèque nationale de France, MS Lat. 7331, is linked to Charles VIII of France (r. 1483–98). Although it is mostly in a style fashionable at c.1450, later additions were made towards the end of the century. One of these is a poem dedicated to an unnamed king, which has been identified as the work of Regnaud le Queux, who held an appointment in the royal household.[30] This in turn suggests that it came into the possession of le Queux, who presented it to the king. However, it contains inscriptions on folios 2v and 61v stating that it was commissioned by one Jehan Roussel, who was probably the same individual who gave another book to Charles VIII. It is thus likely that it was the result of a joint effort. The king apparently gave it away in June 1492, to an anonymous individual in Paris.[31] It is possible that individual was the notorious astrologer Simon de Phares, since the book also bears evidence of having been in the latter's library. However, it was back in the royal library by the eighteenth century, when it was recorded in the catalogue of 1744.[32] If the volume was indeed given to Simon de Phares, then it may have been a show of support by the king, when Simon and his collection of astrological treatises were condemned by the archbishop of Lyon. The case was referred to the Parlement of Paris and the theologians of the university, who impounded all Simon's books from 1491 until 1494. They finally condemned only eleven of them.[33] It is certain that Simon de Phares knew of Fendulus, since he included him in his list of 'Famous Astrologers' and went out of his way to record the hard work done by Fendulus on what Simon called his *Book of the Figures of the Eighth Sphere*.[34] Interest in Fendulus and his vision of the heavens was thus clearly still strong at the end of the fifteenth century.

This copy of the work, like the duc de Berry's, shows evidence of direct connections to Sloane 3983. Its set of images for the Sun originally surrounded a blank page on folio 46r, which was later filled with text in an informal hand. Such a blank page could not have been derived from the de Berry manuscript. Similarly, folio 60r of Lat. 7331 has a horoscope for the birth of Christ. This concept may be rather startling to modern readers, but it was well known in the later middle ages. In this version, it has a reassuring (though misleading) accompanying text stating that it is based on the work of 'Albumasar' as ratified by the theologian Albertus Magnus in his *Speculum astronomiae*. It is closely related to the copy of the same horoscope added on folio 49v of Sloane 3983, again with a reassuring text, although in Lat. 7331 the horoscope is more formally presented. Further attempts to make Fendulus' astrological work an acceptable present for the king are demonstrated by the text on folios 61r and 61v of Lat. 7331, which is a lengthy extract on the winds, taken from the *Aetymologiae* of Isidore of Seville. This is added in a different hand, but is linked to the painted image of the Tower of the Winds, placed on folio 58v. It seems that a text was intended to accompany that image, since the folios between it and the horoscope of Christ are blank, but that plan was not carried out. Once again, the ambitions and claims of both the text and its supporters seem to have created problems for the scribes and artists attempting to produce an impressive version of this mysterious work.

As noted above, the relationship between the Morgan and Sloane volumes is close enough to suggest that the Sloane codex itself may have been in the possession of Abbot Lubertus and used as an exemplar for the duc de Berry's copy. If so, then the horoscope added on folio 49v of the Sloane copy, illustrated with a diagram, may have been included by or for the learned abbot. This addition has the effect of demonstrating the truthfulness and acceptability of astrology, which makes it very interesting that it is not included in the volume presented to the duke. It was, however, copied – and presented more formally – in the volume made for the king. This in turn suggests that the Sloane codex was used as a model for the 'royal' copy, although the latter also received a further addition in the form of the scientific and classical material about the winds. Abbot Lubertus himself had died in 1417, but it is entirely possible that his copy of Fendulus was still available for use as a model by those with powerful connections.

Examination of the handling of the third decan of Aries in Lat. 7331 demonstrates both the similarity to its predecessors and the updating of fashionable details. It also shows that the artists of Lat. 7331 here copied Morgan M785 rather than Sloane 3983. The increased age of the man in red, the clarification of the forge and the transformation of Perseus' sword are all found here just as they are in the duke's book – but not in the Sloane manuscript.

It is, of course, possible that other copies of the work once existed, but have since been lost or destroyed. The fact that authorship of the text was dubious – and that it had no clear title of its own – would make it hard to identify in a library list. As Chapter 8 will show, Charles V of France possessed an impressive library, which included a surprisingly large collection on astronomy and astrology. The combined total for these subjects is over ninety volumes in a book collection that in 1373 comprised nearly 1,200 volumes. Charles' interest in the prediction of future events was strong enough that he also possessed twenty-eight volumes on divination and prognostication, as well as a collection entitled 'Necromancy', treatises on magic rings and two copies of the *Ars notoria*.[35] No astronomical or astrological work in the catalogue is entered under the name of Fendulus, but entry 672 – a volume on astrology – includes an untitled 'new work on astrology'. Perhaps more promising is entry 675, which contained a text 'edited' from the work of 'Albumazar'; several others are simply recorded as the work of 'Albumazar' without titles or details. The library also included a set of books which were once the property of John de Lignières, master of astronomy and astrology in the growing University of Paris, who died c.1335. Among these

Figure 5.11. The final decan of Aries. Paris,
Bibliothèque nationale de France, MS Lat. 7331, f9r.

was a volume with 'figures of the signs of heaven' (829), as well as selected works on judicial astrology.[36] Similarly ambiguous evidence can be found in the 1426 inventory of the Visconti Library in Pavia, which was almost as large as that of Charles V.[37]

Of all the manuscripts discussed in this chapter, it is perhaps the latest copy – with its links to both a royal patron and a famous astrologer – that provides most information about the status and value accorded to Fendulus' work. What the king of France and the royal duke received was a complex and fascinating series of images of the heavens, almost visionary in their intensity and complexity, yet made to refer to contemporary life. This would help viewers, under the guidance of trained astrologers, to understand how these sections of the zodiac could inflect the natures and powers of the planets, and thus nuance their influence on human affairs. However, it is striking that it seems to have been the images of the decans which made most impression on Simon de Phares, since he saw the book as a depiction of the 'figures of the eighth sphere' (i.e. the sphere of the stars, placed higher than the spheres of the planets). On a more ominous note, the idea that specific star groups had power over human beings continued to be both debated and condemned, as the troubles experienced by Simon demonstrate.

This was a topic much discussed in the French royal court, as the country endured succession crises, long-drawn-out wars and disabling royal illness. An important – and brave – contributor was the courtier and scholarly author, Christine de Pizan. Christine's courage was clearly displayed when she gave her support to astrology and astrologers, as well as to women. The beautifully illuminated presentation copies of her works – given by her to her very powerful patrons – will be examined in Chapter 8. But first, it is time to look at evidence for the spread of magical theories and practices beyond royal and imperial courts. At the heart of the next chapter will be a complex text that combines magical concepts, especially the summoning of angels or spirits, with very pious claims. It spread in monasteries and in the new and expanding universities – which were training professionals to work in both courtly and urban environments. This is important since it shows that some forms of magic were finding a place among students and future teachers at the same time as the mysterious work of Fendulus was moving in the circles of rulers and their advisers.

6

THE *ARS NOTORIA*

Ritual Magic for Students and Scholars?

In the case of the influential text usually known as the *Ars notoria*, the manuscript evidence is complex. On the one hand, the number and range of surviving copies testify to its enthusiastic reception, especially among educated individuals with apparently pious motives. On the other, it exists in strikingly different versions, reflecting an awareness of the dangers (both spiritual and legal) facing those who followed its instructions. Unlike the texts discussed in the previous two chapters, the original version of this work is completely anonymous. It is also not possible to identify an early owner. It has taken careful study of medieval library catalogues, surviving manuscripts, theological works and chronicle sources, carried out by several scholars, to establish a set of key points.[1] The first is that it was one of the most well-established works of ritual magic from the early thirteenth century until the beginning of the modern period. The second is that numerous medieval copies were openly recorded, despite the fact that the work was repeatedly condemned by leading legal and theological authorities.[2] Some fifty-six extant manuscripts contain all or significant parts of the treatise.[3] This popularity may be at least partly explained by the fact that it promised to deliver total and (relatively) rapid knowledge of the full range of subjects taught in medieval universities, which normally required many years of hard and expensive study. The third point is that it underwent significant processes of revision, editing, expansion and selection, such that very few surviving manuscripts preserve the same version of this complex treatise.[4]

Nevertheless, one thing is both central to the *Ars notoria* and unmissable to anyone looking at a copy of it: the presence of impressive and very complex diagrammatic images which are both fascinating and challenging. Unlike diagrams in other magical texts, these do not illustrate what the text is seeking to communicate. Nor do they act as models to be replicated in three-dimensional form as pieces of magical equipment.[5] Instead, they are claimed to work almost in the same way as religious icons, or relics in the form of pictures of saints. That is, faithful possession and use of them can offer direct and personal contact with powerful and benign supernatural forces – and ultimately even with God. Also rather like icons and relics, they must be encountered only by someone who is suitably prepared and in a spirit of prayerfulness. What is unique about them, however, is that their visual contents consist of arrangements of symbolic and geometric forms, patterns and symbols, interspersed with 'words', which are frequently unintelligible combinations of letters.[6] In that way, they are in direct contrast to the visual clarity of icons and image-relics. Thus, practitioners who open their minds to receive and imprint these labyrinthine images, while reciting complex verbal formulae and strings of mysterious, almost unpronounceable words and names, are engaging in a significant act of trust. This may perhaps help to explain why so many sought to create their own version of the text, or at least to be given explanations and guidance alongside the mysteries of the oldest version.

Several factors made the *Ars notoria* fundamentally different from other magical texts. First, the rituals framing and shaping usage of the images are presented as extremely pious, and the texts to be recited are identified as prayers. The alien words, names and characters are explained as coming from ancient languages such as Greek, Hebrew and 'Chaldean' and are claimed to preserve both the names of angels and words used to communicate with them. Second, the advantages offered are the relatively virtuous ones of

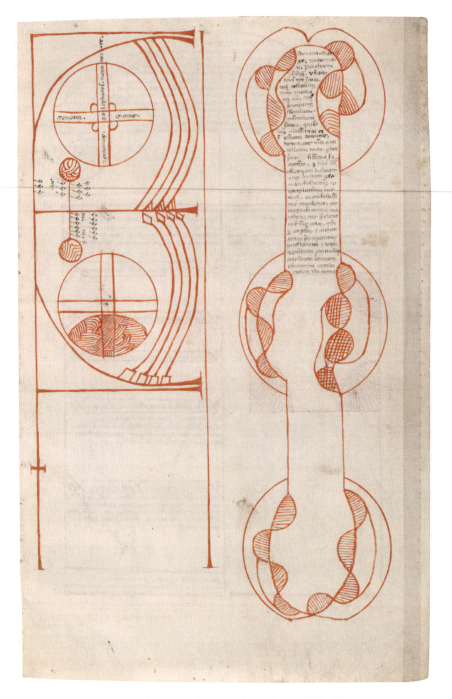

Figure 6.1. Eighth set of *notae*. New Haven, Yale University, Beinecke Rare Book and Manuscript Library, Mellon MS 1, f14r.

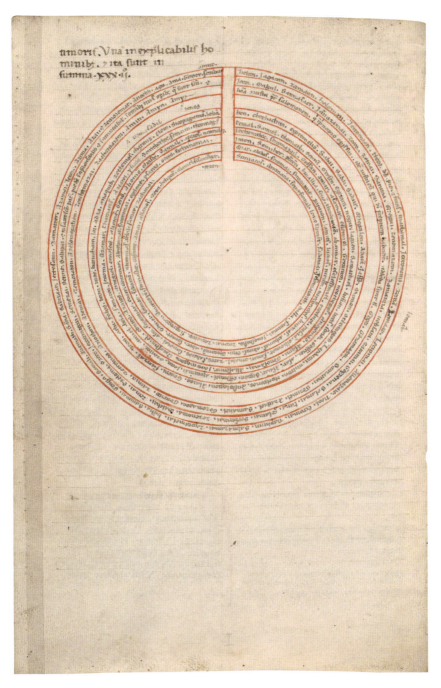

Figure 6.2. First *nota*. New Haven, Yale University,
Beinecke Rare Book and Manuscript Library, Mellon MS 1, f10v.

gaining contact with spiritual beings and full knowledge of the subjects taught in medieval universities. However, such claims cut little ice with thirteenth-century theologians, who saw clear links to things condemned as superstitious and demonic by St Augustine. St Thomas Aquinas was worried enough by the *Ars notoria* to name and condemn it specifically in his *Summa theologiae* – one of the most authoritative summaries of Christian teaching. He dealt with the very serious issue of superstition in Book Two, Part 2, and Question 96 is effectively devoted to the *Ars notoria*. Aquinas' conclusion is wholly negative. The 'art' is both 'unlawful and futile'. It is futile because it cannot deliver what it promises. Still more seriously, its 'signs' are neither understood by humans (like ordinary words and letters) nor sent by God (as sacraments are), and thus are precisely the type of thing that lures humans into contact and compact with demons. Aquinas also dealt with the text's claim to be a record of a divine gift granted to Solomon and made available to readers. He acknowledged that Solomon had indeed received wisdom and knowledge directly from God – but emphasised that this was unique to Solomon, and could not be passed on to others or communicated through 'particular observances'. Aquinas affirms that demons cannot give such knowledge to humans, and states conclusively: 'It is a good thing to acquire knowledge, but it is not good to acquire it by undue means, and it is to this end that the *notory art* tends.'

That may seem a conclusive case for the rejection of the *Ars notoria*, especially as it was echoed by other major theologians. However, the evidence of the number of surviving medieval copies of the work, and the fact that it was copied and owned in religious institutions until the end of the medieval period, contradict this. Moreover, it was translated in the early modern period and also went into print, demonstrating an ongoing – and more widespread – interest. In addition to the copies of all or part of the work itself, discussed by Frank Klaassen, the research of Watson and Fanger has found some twenty copies and extracts of the fourteenth-century adaptation of the original by the French monk John of Morigny.[7] Curiously, John's version, extensively revised under visionary guidance by the Virgin Mary and much less dominated by the characters and signs which troubled Aquinas, was formally condemned and burnt (in 1323), while the 'original' *Ars notoria* never suffered that fate.[8] This condemnation took place at the University of Paris, as is recorded in both the *Grandes Chroniques de France* and the *Cartulary* of the university. No formal record of the terms of the condemnation has been found, though the *Grandes Chroniques* show an awareness that John's work was close to the *Ars notoria* and thus subject to the same problems, despite John's efforts to improve it.

This lack of repeated official prosecution is clearly a problem for historians trying to gauge the impact of the *Ars notoria*. The suggestion by Klaassen, Fanger and Watson that it had a specialised audience and therefore posed little danger to the majority of Christians is convincing. The fact that it was only in the setting of a major university, with a powerful group of theologians, that a new and apparently fashionable version of the work was burnt, clearly supports this.[9] However, it is possible to make further suggestions, at least about the strange characters and words in the original version. These are distinct from the magical 'figures' and 'seals' transmitted in works like *Picatrix*, which were much less like recognisable letters.[10]

The first is that monks studying the multi-layered subject of *computus* (the techniques involved in calculating the date of Easter) were already familiar with the idea of unfamiliar alphabets and signs.[11] Textbooks on *computus* mentioned the various ancient calendars and calculations that lay behind it, including those of the Greeks and Egyptians, and some even gave examples of exotic alphabets. An impressive example is provided by the collection of *computus* materials in a twelfth-century volume probably produced at Thorney Abbey, Cambridgeshire (now Oxford, St John's College, MS 17).

This volume offers a collection of no fewer than thirteen 'alphabets', five of which are versions of the runic *futhorc* (see Figure 3.5). A further four are supposedly derived from ancient sources, and are claimed to be of 'Chaldean', Egyptian, Scythian or Welsh origin in other manuscripts. The final four are cryptic alphabets, intended for use in coded texts.[12] In the Thorney manuscript these alphabets are simply presented as lists of characters, with or without Roman equivalents; but it is worth noting that they are accompanied by complex diagrams and a schematised map.[13]

A second point is that a way may have been prepared for the characters and signs in the *Ars notoria* by the writings of visionaries who attempted to communicate their experiences of contact with the divine. An example is provided by the 'unknown language' and 'unknown alphabet' revealed in visions to the respected abbess Hildegard of Bingen between 1150 and her death in 1179. Hildegard took the step of writing in 1153 to Pope Anastasius IV to inform him that she had miraculously been given these gifts; and they were accepted as of divine origin.[14] Hildegard communicated this to others and incorporated some of her revelatory words into the hymns she composed for her nuns. They were also recorded as significant gifts in formal accounts of her life and miracles. A beautifully written version of both the alphabet and the words (with translations into Latin) is entered into a volume now in Wiesbaden, Hessische Landesbibliothek, MS 2, and folio 464v is shown below (Figure 6.3).[15]

Hildegard was one of the most celebrated visionaries of the late twelfth century and was consulted by both secular and religious leaders across Europe, as her surviving letters show.[16] Her visions also received papal approval; and it is important to note that they included gifts of high-level knowledge that were received instantly and fully understood. Indeed, Hildegard found that she could teach and write on subjects she had never formally studied. As she recorded:

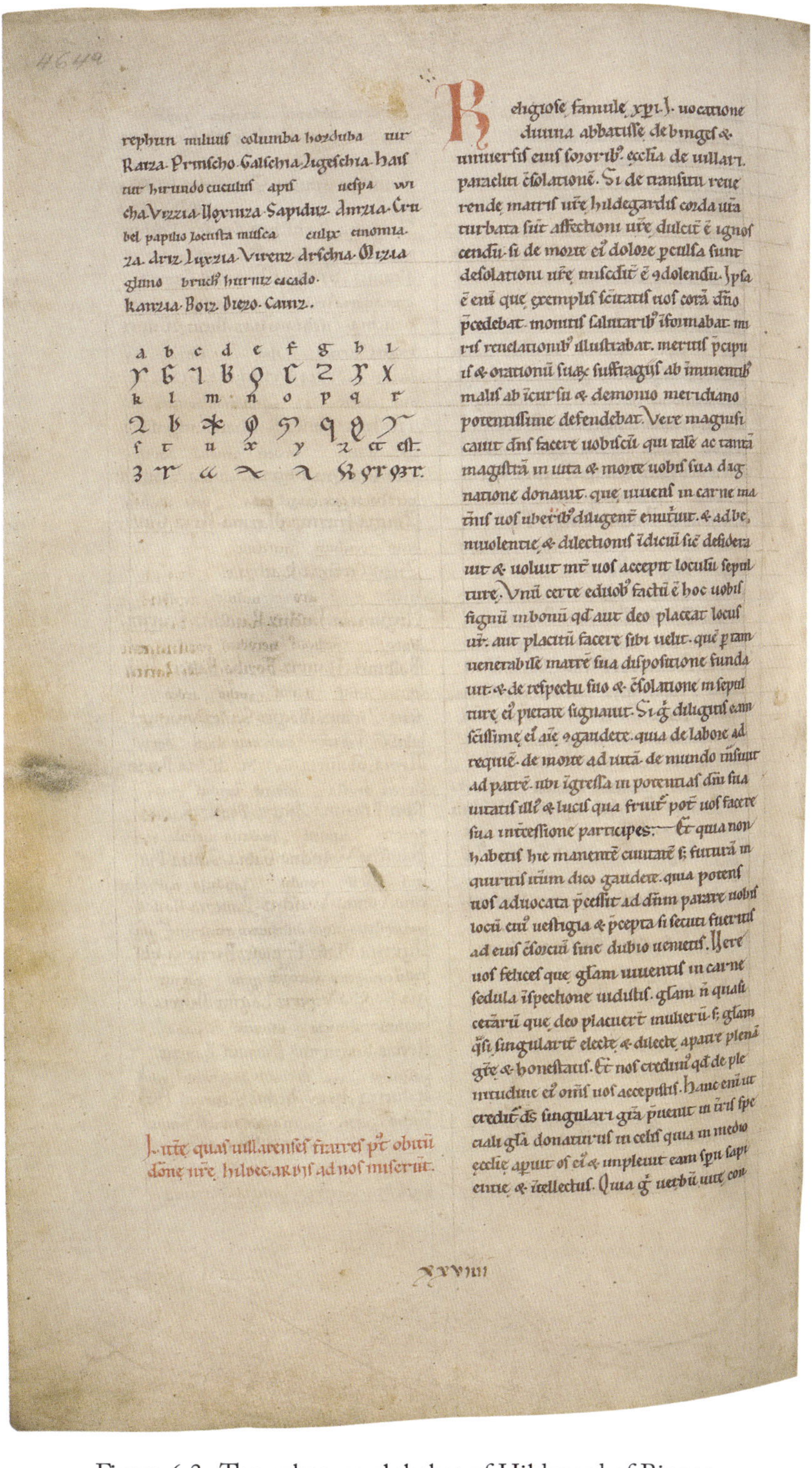

rephun miluus columba horduba tur
Raiza Prunscho Galscina ligescina hail
tur hirundo cucutus apis uespa wi
cha Vrzzia Ilovinza Sapiduz Amzia Cru
bel papilio locusta musca culpr enomia
za driz Luzzia Virenz drscina Orzia
glmio bruck hurntz cacado.
kanzia Boiz Biezo Carnz.

a b c d e f g h i
k l m n o p q r
s t u x y z et cet.

R religiose famule xpi l. uocatione
diuina abbatisse de binges &
universis eius sororib; ecclia de uillari.
paracliti csolatione. Si de transitu reue
rende matris ure hildegardis corda ura
turbata sint affectioni ure dulcit e igno
cendu. si de morte ei dolore pensita sunt
desolationi ure miscetur e adolendu. Ipsa
e eni que exemplis scitatis uos cora dno
pcedebat. monitis salutarib; isormabat in
ris renelationib; illustrabat. meritis pcipu
is & orationu suaq; suffragiis ab immentis
malis ab icursu & demonio meridiano
potentissime defendebat. Vere magisti
cauit dns facere uobiscu qui tale ac tanti
magistra in uita & morte uobis sua dig
natione donauit. que uiuens in carne ma
tris uos uberti; diligent enutriut. & ad ue
niuolentie & dilectionis idiciu sic desidera
uit & uoluit mt uos accepti loculu sepul
ture. Vnu certe edtiob; factu e hoc uobis
signu in bonu qd aut deo placeat locus
ur. aut placitu facere sibi uelit. que p tam
uenerabile matre sua dispositione funda
uit & de respectu suo & csolatione in sepul
ture ei pietatis signauit. Si q; diligitis eam
scissime ei aie agaudere. quia de labore ad
requie. de morte ad uita. de mundo insuir
ad patre. ubi igressa in potentias dni sua
uitatis ille & lucis quia fruit pot uos facere
sua intcessione participes. — Et quia non
habetis hic manente cuitate s; futura in
quiritis itim dico gaudere. quia potens
uos aduocata pcessit ad dnm patre nobis
locu cui uestigia & pcepta si secuti fueritis
ad eius csoiciu sine dubio uenietis. Ifere
uos felices que glam uiuentis in carne
sedula ispectione uidutis. glam n quali
cetaru que deo placuerit mulieru s; glam
qsi singulari; electe & dilecte apatre plena
gte & honestatis. Et nos credim qd de ple
nitudine ei omis uos accepitis. hanc eni ut
credit dcs singulari gra puenit in uiri spe
ciali gla donatu uis in celis quia in medio
ecclie apuit os ei q; impleuit eam spu sapi
entie & itellectui. Quia q; uerbu uite con

l nie quas uillarenses fratres pt obuii
done ure hildegarois ad nos miserut.

XXVIII

Figure 6.3. The unknown alphabet of Hildegard of Bingen.
Wiesbaden, Hochschul-und Landesbibliothek
RheinMain, MS 2, f464v.

I understood the writings of the prophets, the evangelists and other saints, and those of certain philosophers, although I had not had any human instruction; and I expounded some of their writings . . . But also I composed and sang songs with melodies in praise of God and the saints, again without any human instruction, although I had never learned neums [notation] or singing.[17]

In other words, Hildegard was given instant mastery of, at the least, the arts of Grammar, Theology, Philosophy and Music, which would normally require years of study. Moreover, Hildegard's visionary experiences were often expressed in visual terms, which were communicated to assistants and turned into striking images, as well as verbal accounts. The original manuscript containing these has been lost since the Second World War and probably destroyed, but an accurate copy was made at Eibingen Abbey before that. The range of both the visions and the images is very wide, but some of the forms used to structure the visual versions can be compared to those found in the *Ars notoria* (see Figures 6.4a and 6.4b).

This is not to suggest any direct copying of Hildegard's images on the part of the creators of early versions of the *Ars notoria*, but rather a possible source of support for those who wished to see it as both orthodox and potentially effective. Nevertheless, it must be said that the surviving early copies of the *Ars notoria*, which seem to originate in university circles (as would be expected), hardly offer an effortless source of knowledge and inspiration. The copy now in Yale University Library (shown in Figures 6.1 and 6.2) is accepted as the earliest known version, and it has been suggested that it was produced in the setting of the University of Bologna.[18] It can be described as a copy intended for a university master, since to use it requires considerable preparation, experience and confidence.

auī acutaſ aureſ inꝛioreſ ītellec͛ꝯ vīīī. De differentia ꝫ unitate triū pſonaꝛ.
habet. hic ī ardente amore ſpecul͛ x. De trib͛ ſimilitudinib͛ ad trinitate.
mei aduerba hec anhelet. ꝫ ea ī xī. Verba de libro regū ad eande reo).
ꝓſcientia animi ſui conſcribat.

Explicit ſexta uiſio tertie partiſ. Capitula ſeptime uiſi
oniſ tertie partiſ.

Q d ineffabiliſ trinitaſ iſt ī ne t͛pꝙ
declarata ſimplici ꝫ humili
a fidelib͛ credenda ꝫ colenda
ē. ne q̅ſ pluſ inueſtiganſ quā
oportet: q̅a ꝓphendi ñ poteſt.
ī deteri cadat.

II Q d ī ſangu͞e xp̄i munduſ ſal
uatuſ ē. ꝫ cultuſ ſc̄e trinitatiſ
manifeſtiſſime declaratꝫ ē. ipſa
tam nulli intelleꝯui patet.

III Q d ineffabiliſ trinitaſ oniſ creatu
re aptiſſime ī ipro ꝫ poteſtate
apparet. excepuſ ī creduliſ cor
dib͛. cuncta tam uelut ī eadeī
gladiuſ penetrat. § ſuccidit.

IIII. Qui xp̄iano pp̄lo catholice fidi
ī arduitate ī fidelitaſ aduer
ſant. hoſ diuinitaſ ī ꝓfuſione͞

V Q d diuinitaſ iactantia iudaici
pp̄li deicit.

VI Q d diabolicū ſciſma gentiliꝯ ipſi
a d͞o abſceſu uadit ī pdicione.

VII Parabola ad eande re.
VIII Verba iohīſ ad eande re.

Figure 6.4a. *Scivias* manuscript, plate 28: Hildegard's vision of the effects of
sin. Eibingen, Abtei St Hildegard, Cod. 1.

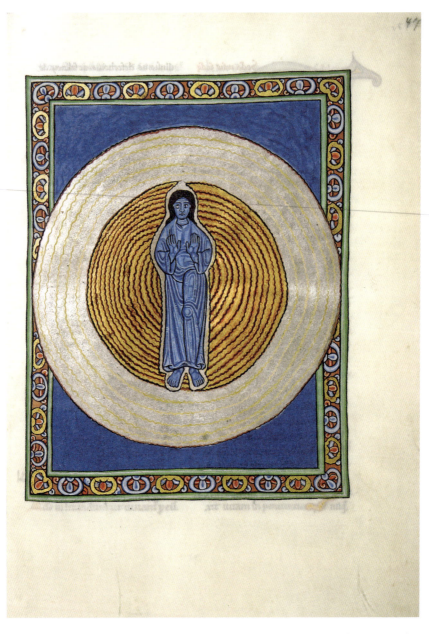

Figure 6.4b. *Scivias* manuscript, plate 11: Hildegard's vision of a quiet light encircling a man in sapphire blue. Eibingen, Abtei St Hildegard, Cod. 1.

YALE UNIVERSITY, BEINECKE RARE BOOK AND MANUSCRIPT LIBRARY, MELLON MS 1

This manuscript now has neither an original flyleaf nor a title page. It opens on folio 1 with a grand initial 'I' in red and blue, running down the left margin. This is the first letter of the announcement that 'here begins the prologue'. The grandeur continues, since all this introductory section is written in impressive red ink. This very first paragraph establishes the pedigree of the work, announcing not only that what follows is a 'book of divine wisdom' revealed via an angel to Solomon, but also that it has passed through the hands of 'Apollonius the great', who gave the first part the name of *Golden Flowers*. Its relevance for its university audience is made clear by the rather surprising, first-person opening of the next section: 'I, hailed as Master of the Arts'. The anonymous master (who may or may not be Apollonius himself) goes on to provide testimony for the efficacy and power of the work, together with an outline of what it can offer. This figure provides the 'voice' delivering definitions and instructions embedded in the text that follows. Consequently, the work has up to three powerful figures behind it – Solomon, Apollonius and (perhaps) the more modern master.

Guidance and instruction are certainly needed, as the reader is plunged into a complex text, which, it is directly stated, operates in ways that cannot be simply understood by human reason and that require the correct recitation of lists of 'unknown' names. Statements on the origins and power of the treatise's contents are interspersed among texts to be recited as prayers (orations) and lists of mysterious words and names. Key terms are: words (*verba*); figures; *notae*; and (spoken) prayers (*orationes*). No straightforward definitions of these are given. To make things more complicated, the order in which passages of different types of text are to be used is not always entirely clear, meaning that the practitioner needs to make a number of decisions as to when the strange words and names are to be recited in relation to the prayers. Even the correct pronunciation of the names would be a matter of concern, as a short quotation from the first 'prayer' will show: 'Phos, Megale, Patir, Ymos, Ebel, Eber, Helioth, Gezei, Salatial, Sadim, Helgyo, Megis, Micton, Esel, Gecor, Granal, Semaranxai, Gelsemana, Arasamion, Sale, Patir, Agion, Atnas, aminb.'[19]

Nevertheless, it is clear that materials for the improvement of the would-be practitioner come first. These lead to the acquisition of the 'general' gifts of memory, intelligence and eloquence, drawing upon the 'figures or prayers called *triumphales* by Solomon'. With this stage achieved, the practitioner can move on to the 'specials'. This, however, cannot be rushed, since the 'triumphals' must be repeated on eight days within a complete lunar month. The first performance should take place on the fourth day, and is repeated on the eighth, twelfth, sixteenth, twentieth, twenty-fourth, twenty-eighth and thirtieth days. It is worth pointing out that the use of lunar days would be familiar to clerics trained in the liturgical calendar of the Church, as well as to astronomers, so this would not in itself appear sinister, despite Aquinas' reservations. The 'specials' begin with the three fundamental arts of the university syllabus: Grammar, Eloquence/Rhetoric and Dialectic, which were collectively known as the *trivium*. Interestingly, the text devotes a good deal of space to these, despite their relatively low status. They are then more briefly followed by materials for a rather unorthodox version of the more advanced subjects of study. As given here, they are Philosophy, Medicine, Music, Geometry, Mathematics and Theology.[20]

The eightfold recitation of the 'triumphals' gives a month-long structure to the core ritual of the *Ars notoria*, linked to the days of the Moon (i.e. the days of a lunar month, starting at new moon). These could be identified using a standard liturgical calendar and did not require direct observation. These recitations could be interwoven with performance of the 'specials', although the linking of all this to the images is not yet entirely clear. Logically enough, the 'specials' are followed by texts to ensure success in the combined usage of the prayers, names and images. However, anyone expecting to progress immediately to the images will be surprised. What in fact follows is a set of additional materials for mastery of theology, suggesting perhaps that this was a particular interest of early users. This group is then further followed by a short text named by modern editors as the *Ars nova* or *New Art*. It offers practical help for those who cannot dedicate such long periods of time as are required by the original *Ars notoria*, since it consists of a group of new prayers. Helpfully, these can simply be recited immediately before using the images. However, they can also be used in addition to those already given. The latter point is important since it helps to explain why these more straightforward texts are still followed by prayers to be recited while looking deep into the images, and by further instructions concerning the *notae*. It is thus very likely that this manuscript already represents a fairly advanced stage in the evolution of the *Ars notoria*, since additions and improvements appear to have been inserted.

As this description makes clear, the images are given only after these sets of complex textual materials and instructions. There is a logic in this, since the practitioner must be trained and prepared before attempting to use the images. However, it will also be clear that the book is not intended to be simply read through as presented. Instead, the practitioner is required to assimilate not only the texts themselves, and the instructions for their use, but also the specific combinations of text and image appropriate for each individual art. For this reason, it seems likely that the copying out of a leaflet containing the textual elements needed, in order, and with projected dates, would be very useful. Preliminary study, and ideally memorisation, of these would free the practitioner to focus exclusively on the folio containing the requisite image, rather than having to consult several different folios simultaneously. This necessity for individual creativity would also help to explain the variations found between surviving copies of the overall work. As the illustrations above show, the images themselves require a high level of technical and artistic skill, if their components are to be correctly represented and placed, so use of the more formal, professional version of these would be reassuring. The practitioner, even once thus prepared, would also need to have the means and the time to plan the performance of each ritual across at least a full lunar month. This is clearly a lot less than would be required for standard university study of the same subject, but is still some way from the sort of instant revelation described by Hildegard of Bingen.

One thing that is not spelt out, but appears implicit in the structure of the treatise, is that the practitioner will gain expertise in the process of working through the rituals for each art. The first, Grammar, has three images, which are to be looked deeply into (the Latin word is *inspectio*) in a specified order and separately from the recitations of the preliminary prayers. These 'lookings' are to take place in the evenings, with the first image being the focus of days one to fourteen of the chosen month. The words incorporated into the image are to be read out or recited twenty-four times, and books of grammar are to be opened and briefly looked at during this time. Then the image itself is to be looked into, piously and solemnly, twelve times. From then until the twenty-eighth day, the second figure is to be added, and the work required increases. Each figure is looked at twenty times, and there are to be thirty recitations of the texts. On the last two days, all three figures are to be used, looked into twelve times each, with thirty-seven recitations of the prayers.

However, subsequent rituals are somewhat simpler, implying that the practitioner will build on what has already been achieved. Dialectic has only two figures, both of which are to be looked into every evening for a month, with twenty recitations of the prayers. There is another echo of the church calendar here, as the figures are not to be looked at on the so-called Egyptian days. These were unfortunate days, which were transmitted with varying details; but the most common version identified two or three days each solar month, usually towards the start and end of the month. Again, the prescription is not explained, but the idea would be familiar to clerics, even though theologians disapproved of this concept by the thirteenth century. Rather counterintuitively, the more advanced arts have the simplest rituals, while also leaving most to the discretion of the practitioner. The figures for each of the subjects of the *quadrivium* – Astronomy, Mathematics, Music and Geometry – are only to be looked into on the first four days of the month, and the number of recitations of the prayers is left up to the practitioner.

Philosophy has a unique procedure, having no fewer than seven figures, which are to be looked into on the seventh, seventeenth and twenty-seventh days of the month. The prayers for each are to be recited seven times. In a similar way, Theology, the highest subject of all, has a special procedure. What is rather more surprising is that the arts of magic themselves are incorporated into the treatise and its rituals, under the heading of 'select' or 'exceptional' arts. These are dealt with rather awkwardly, mostly by looking into the figures or images called 'generals', although Mellon 1 also has an image labelled as suitable.

The images are to be looked into ten times every day, with recitations of the orations. It is also assumed that the practitioner will have copies of appropriate books to look at briefly during the process. All this makes it possible that 'Magic' was added as something of an afterthought, which would again support the possibility that the Mellon manuscript already represents a relatively advanced stage in the development of the text. However, it is also possible that the relatively small space given to this subject would make it less conspicuous if the book were to be examined.

JOHN OF MORIGNY, BRIDGET AND THE *ARS NOTORIA*

An expert user of the *Ars notoria* who became convinced that it was corrupt and even demonic was Brother John, a fourteenth-century

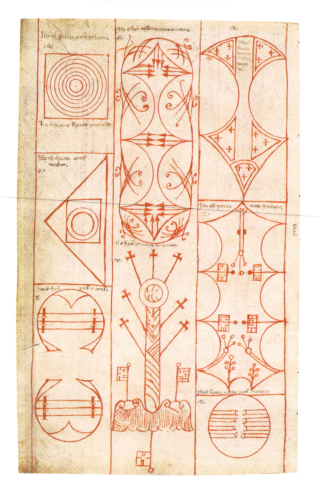

Figure 6.5. Figure for 'all the Exceptionals'
(lower image in central column). New Haven,
Yale University, Beinecke Rare Book and
Manuscript Library, Mellon MS 1, f16v.

monk of the abbey of Morigny, near Étampes. John's account of his
introduction to the text, his powerful attraction to it, and the terrifying experiences he underwent while using it, is unique and
important.[21] This account was given in several chapters of John's
own, visionary work, which went under varying medieval titles but is
now known as *Liber florum celestis doctrine*. John's susceptibility to

visionary experiences seems to have begun early in his life, since he records that the first happened when he was about thirteen years old, and it left him with the conviction that the Virgin was his special protector. This experience would have taken place at the end of the thirteenth century, when John seems to have been a schoolboy in Chartres, perhaps as a choirboy. It was no simple vision of the Virgin, but rather began with the terrifying apparition of a huge, demonic figure. John fled in his vision into the cathedral and was saved by the intervention of the Virgin. Understandably the experience left a lasting impression on John and seems to have encouraged him in the belief that his later visions were worthy of being written down and preached about, after he had become a monk. It is clear from John's account that the great attraction of the *Ars notoria* for him was its promise of rapid access to advanced scholarly knowledge. He recounts his continuing wish to study, and how he was lent a copy of a book of necromancy by 'a certain cleric'. No details are given of this volume, but it was very attractive to John, who copied much of it and wanted more. This led to an encounter with a 'medical expert' from Lombardy, who informed John that what he needed was the *Ars notoria*, and that a copy of it was to be found within the walls of the school (at Orléans) where John had been sent. These details provide important evidence of the liminal status of works of ritual magic. They were recognised as dangerous, and were far from being officially approved, and yet were well known, owned and recommended by educated individuals, including monks and clerics.

For John, the *Ars notoria* was dangerous not in some abstract way, but very directly. He confesses that it struck him at first as beautiful and holy, and offering miraculous gifts rather than demonic temptations. However, it became apparent that the book was utterly deceptive, a work of the Devil, a 'sick pleasure' which was actually fatally poisonous to the soul, not only the body. Even within the condemnation, the nature of the attraction is revealed when John says that, without the *Ars notoria*, attempts at necromancy will fail (Chapter 13).

At the same time, he says that 'it cannot be understood by anyone' and 'the more you study it the more obscure it becomes'. The very first experiment with a ritual from the book led to a vision of John's mother's house, lit by moonlight. Like the introductory vision, this one was terrifying, and involved the slow apparition of a huge demon, which transformed into the likeness of a holy man dressed in black and wearing a black crown. John did not fear that the holy man would kill him and the vision ended when he woke up, without needing the intervention of the Virgin – again emphasising the ambiguous status accorded to the book. In fact, when John tried to understand the vision he made use of a prayer from the *Ars* itself (Chapter 14) seeing no harm in it, and immediately experienced another vision. It was this new vision that contained a warning, and that led John to have doubts about his beloved book.

On a more prosaic level, John also records that the *Ars notoria* was especially attractive because he could not afford the full range of books needed for his studies, nor the full series of lectures. He seems to have abandoned his official course, wishing to take advantage of the promise of knowledge of all the sciences, and devoted himself to working with the *Ars*. His reward was that he believed he had worked out how to follow the rituals and make them work – and he followed one (unnamed) ritual for a whole lunar month, ending correctly on the night before the new moon. That night John experienced another disturbingly ambiguous vision, in which three figures (whom he wrongly thought to be the Trinity) ordered him to continue using the book for a further eight days. Further troubling visions ensued, in which arrogant figures demanded worship – something that was clearly against true religion. Gradually it became clear to John that the obscurity of the book hid invocations to demons within what appeared to be beautiful prayers. And yet, like any addict, John found it almost impossible to give up the *Ars notoria* completely.

This was partly because it helped him to learn many of the recognised forms of magic and divination. It was only when the visions became directly life-threatening that the Virgin, with St John the Evangelist as her intermediary, once again appeared and made it clear that, as John says, 'the *Ars notoria* was deeply evil' (Chapter 21). This revelation, however, only inspired John to go back to direct study of necromancy: he even boasts of his success in composing a new work of his own, as well as of making the magical Rings of Solomon. Further visionary experiences were required before he resolved to give up necromancy completely. The next chapter reveals just how strong was John's attachment to the use of ritual magic in the form of prayer, meditation and subsequent visions. Here both the Virgin herself and a company of angels are invoked to support the assertion that John next composed prayers and accounts of his visions. He then went on to put together a new, short and clear collection of thirty prayers which he believed would lead to a genuine knowledge of all that the *Ars notoria* had falsely promised. John was to work on, and expand, this for a considerable length of time, and it gained a wide readership in its own right both before and after being condemned in Paris.

The case of John of Morigny provides valuable evidence of the powerful attraction exercised by the *Ars notoria* over ambitious scholars whose means were limited. The very evident tension between a desire to believe the mysterious texts and images to be beneficial and devout, on the one hand, and a fear that its mysteries hid demonic traps on the other hand is also unsurprising. Perhaps less predictable is the reality of the powerful and terrifying experiences that John underwent as a result of carrying out the rituals and meditations. These may be linked to the fact that he had already undergone visionary experiences, but that does not rule out the possibility that others were also strongly affected. What is certain is that John's revelations were taken very seriously both within his own monastery and well beyond it.

However, John's account is perhaps still more important and unexpected for a different reason. This is that he records giving instruction in using the *Ars notoria* to his younger sister, Bridget, whom he also taught to read Latin. This took place when Bridget was about fifteen – which John describes as a very late age to begin such studies. It seems that John wished to speed up Bridget's learning, to compensate for her late start, and so used the *Ars* as the basic text for her to learn the alphabet and begin to pick out words. Bridget's progress was exceptionally fast and impressive, giving evidence for John of the effectiveness of the *Ars*, and at some point she began to use its full rituals for herself. However, like her brother, Bridget paid a price in the form of terrifying nocturnal visions of a demonic creature which oppressed and threatened her. The spirit was defeated by prayer and an oath to renounce the *Ars notoria*, which led to visions of both God and the Virgin, and ultimately to entry into a religious house which John calls *Rosetum*, even though the family had no connections there. This seems to have been a well-regarded institution, since John is very pleased to have secured for Bridget the third vacancy to become available, whenever that might be. Knowledge of the *Ars notoria*, presumably accompanied by condemnation of it, would thus have been taken to that religious house, as well as to Morigny. Indeed, its attraction to members of religious communities seems to have been strong, since John was asked to expound it to a monk of the Cistercian house of Fontainejean, as well as to discuss it with members of his own abbey. A cycle of prayer and meditation, focused by deep 'inspection' of images and timed in accordance with the Church's luni-solar calendar, would obviously fit well into the life of regular clergy.

THE REVISED EDITIONS OF THE
ARS NOTORIA

Both the subject matter of the *Ars notoria* and the case of John of Morigny suggest that its readers and users were primarily individuals actively engaged in academic studies. This is further reinforced by the suggestion that at least some copies of the work originated in Bologna, where the university claimed an origin in the eleventh century and was granted a charter in the twelfth. A manuscript now Turin, Biblioteca Nazionale, MS E V 13, and probably produced in the second half of the thirteenth century, may come from such a centre.[22] However, its script and illuminations have also been compared to those in manuscripts copied by professional scribes based in Oxford; and it was edited and annotated in c.1300 by an English scribe.[23] The designer of the *notae* in this manuscript has worked hard to make them appear both more familiar and less confusing. These characteristics can be seen in several images, with the first being its version of the first *nota* for Grammar (on folio 1v). The narrow, concentric circles, filled with mysterious words and names, are still the dominant element of the design; but in this unique version they are surrounded by images more usually seen in psalters and prayerbooks. The open centre of the circles is filled with an ink drawing of the Crucifixion, while the corners have drawings of the winged creatures that were long-established symbols of the four evangelists: the winged man of St Matthew, the lion of St Mark, the ox of St Luke and the eagle of St John. To remove all doubt as to their identity, each is holding an open scroll on which is written a well-known phrase from his gospel. Fear that such images might open the way to demonic temptation is thus greatly reduced.

This 'English/Turin' edition of the *Ars notoria* is unique in ways which suggest that it is the result of a special commission. It not only proclaims the spiritual safety of the main text, but also incorporates materials dedicated to a simple form of medical practice. Its first owner was clearly wealthy, as is shown by the size and professional appearance of the manuscript. Moreover, the original contents of the manuscript (which has only thirty-two folios) do not suggest that he desired a compilation of magical texts, since they consist only of this unique version of the *Ars notoria*. Equally, however, the 'medical' content added into this manuscript is of a limited and general nature, suggesting that this was not the book of a trained physician. The added medical material is found on folios 22r to 23v, and has been analysed by Julien Véronèse.[24] On folios 22r and 23v are two unusual *notae* for acquisition of medical knowledge (*Scientia medicinalis*) which may have been created especially for this manuscript. On folio 22v is a diagrammatic image, headed *Nota pronosticationis boni et mali*, which at first sight fits well with the other *notae*. However, it is in fact a version of the long-established and widely circulated prognostic also known as the 'Sphere of Petosiris' or 'Sphere of Apuleius'. This appears in Latin versions in Carolingian manuscripts and was copied in monastic scriptoria across tenth- and eleventh-century Europe, including in England.[25] It remained popular throughout the medieval period, perhaps partly because its use did not require any medical training.[26] The prognosis was based upon ascribing numerical values to the patient's name and the date of onset of the illness (usually), and then following a simple mathematical process to identify a final number. The position of this number in the circular diagram provided the prognosis of whether the outcome of the illness would be positive or negative, and over what time scale. The procedure could, of course, be used to provide answers to other types of problem; but the flanking of this version by *notae* for Medicine provides a clear context. A legendary account of the origin of the device, and information on its use, is provided by the text on folio 23r, which is presented as an authenticating letter.

This unique 'edition' of the *Ars notoria* was presumably not intended to be a model for later copies, and no such copies have been found. However, it strongly suggests that users of this text felt free to treat it as a sort of toolkit of materials, to be drawn upon to suit their needs, and also as a source of positive benefits. The Turin manuscript also provides evidence of long-lasting interest in the overall work, since its fourteenth-century additions and annotations are designed to bring it more closely into line with other copies. The editor, who used an English style of script and thus shows that the manuscript was in English hands at this time, drew upon several other versions of the *Ars notoria* in his work – perhaps as many as three. The identity and exact location of the editor are unknown, but the fact that he had access to so many copies of this supposedly forbidden work is striking. It is also in line with the evidence provided by the book catalogue of St Augustine's Abbey, Canterbury. This catalogue and the surviving manuscripts with magical contents have been extensively discussed by Sophie Page.[27] By the fifteenth century, St Augustine's appears to have owned three copies of the *Ars*, two given by the fourteenth-century monks John of London and Michael of Northgate, who both donated rather impressive collections of magical texts. Sadly, neither of these copies is known to survive. The third copy, which has the owner-inscription of Simon Maydestone, is now Oxford, Bodleian Library, Bodley MS 951. The manuscript does not have a St Augustine's ownership mark, but it is possible that this has been lost. Certainly, the text of the *Ars notoria* is incomplete. However, much survives and it is clear that this is a copy of yet another edition, amounting to a new version, of the *Ars*. What is immediately striking about this new version, discussed in detail by Véronèse, is that it goes much further in rendering the *Ars notoria* safer and less potentially threatening. The manuscript is very similar to one now Paris, Bibliothèque nationale de France, MS Lat. 9336, which was probably produced in northern Italy (perhaps Bologna) in the fourteenth century, but whose owners are unknown.[28]

In this version, the *notae* retain only their geometric construction and have shed their more challenging visual elements. The lists of strange words and names remain as integral parts of the ritual, but its prayerful nature is emphasised very strongly. The first *nota* is found on folio 18r and is headed, in red (underlined by a later user), 'This is the first Figure of the art of Grammar.' Its structure is by now familiar.

Figure 6.6. The first figure of the art of Grammar. Paris,
Bibliothèque nationale de France, MS Lat. 9336, f18r.

However, despite the recognisable structure, it is also clear that a lot of additional material has been introduced into what is here called a 'figure'. Small versions of strange characters have been introduced into the central circle and are echoed in the truncated wedge set below the circle and reaching onto the very bottom of the folio. Anxiety about their nature is immediately contradicted by the presence on the folio of no fewer than seven angels, all of whom hold small crosses. They are all shown in three-quarter view, looking at – and pointing to – the various textual and symbolic contents of this page and its predecessor. Even if they are not necessarily the spirits invoked by the mysterious names to be recited while gazing at the figure, they clearly act as guarantors of its entirely Christian nature. Moreover, the texts added around the figure give helpfully full directions to the user, providing both short prayers to be said in conjunction with the contents of the figure and instructions about the order in which the various components are to be used.

The preceding folios, containing the preliminary 'prayers' as usual, have taken on an appearance that will have been very familiar to university academics and scholarly monks – that of a major religious or canon law text, complete with technical commentaries.

In this comparison, the first image (Figure 6.7a) is of one of the early pages of the revised version of the *Ars notoria*, while the one on the right (Figure 6.7b) is from a collection of canon law texts (datable to c.1200). In both cases, the main text is set out in the two central columns, with explanatory comments and guidance arranged in sections on all four sides. The use of red for short headings, marking breaks between sections of text, is also parallel. The curious 'monstrous' creature in the illuminated initial of the canon law text was standard for decorated manuscripts of the time and helps to demonstrate that the presence of apparently incongruous imagery on pages of entirely orthodox texts was unproblematic. It is not possible to know what the large initials in the *Ars notoria*

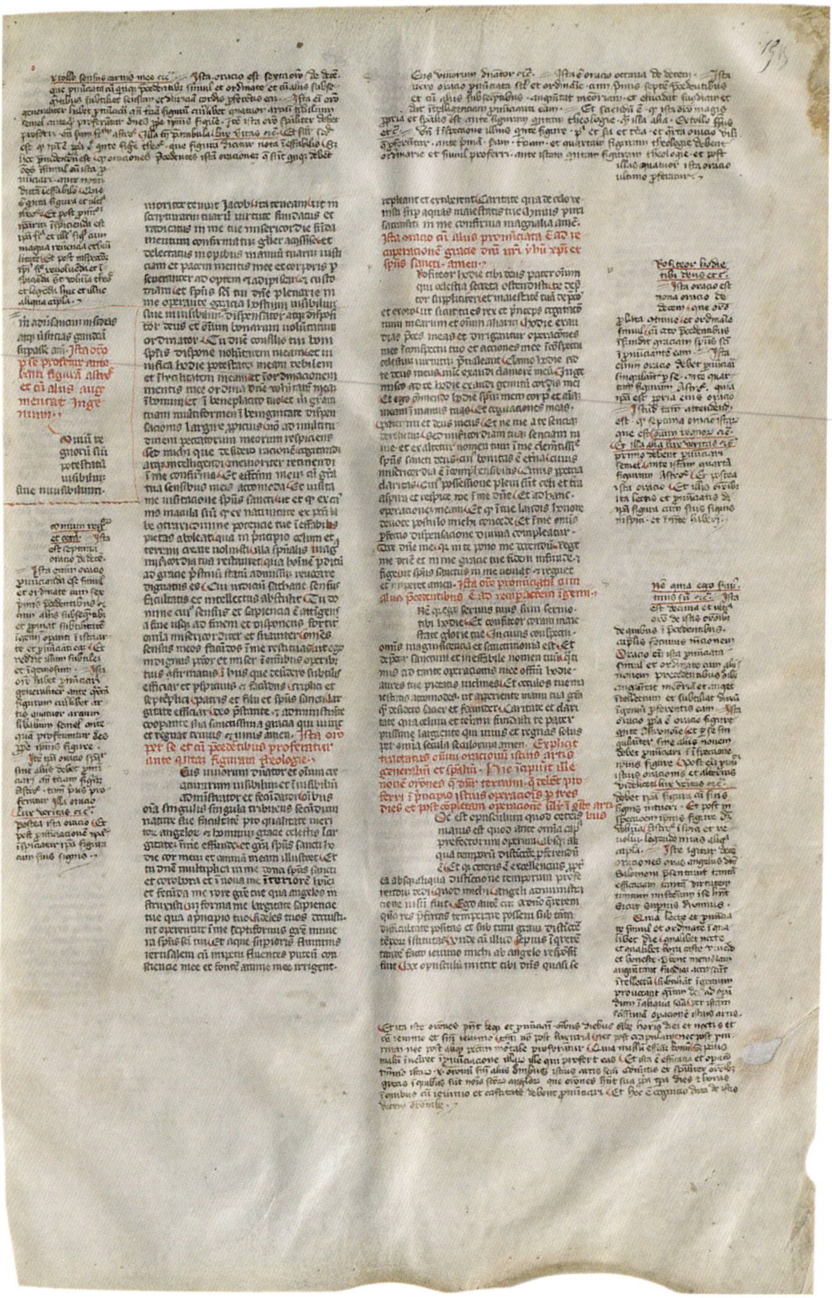

Figure 6.7a. Preliminary prayers of the *Ars notoria*. Paris, Bibliothèque nationale de France, MS Lat. 9336, f15r.

Figure 6.7b. Gratian's *Decretum* (with glosses); part of Causa 1. Paris, Bibliothèque nationale de France, MS Lat. 14316, f75r.

manuscript would have looked like, since they were never painted into the spaces provided for them.

The similarity to a book designed for lengthy study, memorisation and professional use is significant in another way – namely that this version of the *Ars notoria* also required the user to engage with its contents over long periods of time. As noted above, the text comes complete with its own 'user's manual', and this guidance makes it clear that full performance of the rituals could take at least as long as working through a textbook. The first step was to recite the prayer given on folio 1r, followed by the subsequent group of prayers, and rounded off with another recitation of the first prayer. If correctly performed, this should grant a vision giving the user licence to proceed to the next stage. However, it was equally necessary for the practitioner to prepare him or herself through following guidance concerning diet, chastity, meditation and prayer – things which again fitted more easily into a monastic way of life than many others. For those who succeeded in both aspects of the preparation, the next step was the memorisation, recitation and internalisation of the 'general' prayers and *notae*. These are divided into groups, each to be accompanied by a specific and lengthy ritual. Working through each stage would grant the successful student the grace of Christ in pursuing knowledge; outstanding gifts of memory, eloquence and intelligence; and more specialised gifts, based on choice of prayers and names.

All this sounds like a blending of monastic ideals and practices with university study; but parts of the commentary make it clear that the user might intend to practise as a magician. Certain prayers are identified as suitable for use on their own, in order to gain knowledge of future events, such as the outcome of illness or the resolution (or not) of disputes. In other words, portions of this text could be used in much the same ways as the prognostic and divinatory devices collected by Matthew Paris (see Chapter 3). However, it is unlikely that the owner of Lat. 9336 would have stopped at this stage, since triumph over all branches of formal knowledge is again promised by the following sections. Next come the 'undergraduate' subjects of Grammar, Rhetoric and Logic, each with carefully identified sets of prayers and *notae*. Red rubrics point the reader to the sections of text giving the numbers of *notae* to look for when internalising each subject (three, two and four, respectively). There is also a long set of eighteen prayers which must be recited in specified ways before the *notae* can be used. As might be expected, all this is followed by the materials dedicated to the more advanced subjects – usually specified as Mathematics, Geometry, Music and Astronomy. The ambitious user need not stop there, since attention is also given to Medicine and Theology.

The notes of guidance explain that suitable lunar months should be chosen to embark on each subject, and they help in making that choice. They also make it clear that anyone following this full version of the ritual needs to be prepared to devote considerable time and resources to it – especially since they should preferably follow the completion of at least the preliminary stage outlined above.[29] First comes the purging of sin through going to confession and carrying out the assigned penance. Second comes a more physical purging, reminiscent of early medieval practices for the casting-out of illness. This involves the collection of four olive, bay or vine leaves and the beginning of a process of drying them in a clean glass vessel. Even this stage required careful calculation, since the day in question must fall fourteen days before a Sunday, which should ideally be the last day of a lunar month (i.e. fall at the dark of the moon). To priests and monks trained in the use of a luni-solar calendar in order to calculate the date of Easter Sunday in any given year, such specifications will have been familiar; but it is worth observing that patience might be needed if a new lunar month beginning on a Monday was to be awaited. In order to be ready for the Monday of the new moon, the practitioner should take the semi-dried leaves from the glass vessel on the Friday preceding the crucial Sunday, and soak them in a solution of saffron and rose water. Mysterious names should then be written on the underside of each leaf, before they are immersed in sequence in pure water. The names should be rubbed so that they dissolve into the water and this solution should be drunk, without finishing it, and with brief prayers recited between sips. The rest of the day is occupied with carefully timed recitations of sets of the prayers already given. During the whole day a strict fast on bread and water should be observed. Similarly precise and demanding prescriptions are given for Saturday (when the water containing the mysterious names should be finished). On Sunday there is no drinking of the solution, and the fast is replaced by the eating of any preferred foods, once the recitation of prayers has been completed. During this day, acts of charity should be performed. If some days remain before the new moon and the start of the chosen lunar month, then the user should recite the prayers repeatedly while maintaining physical and spiritual purity – although students should attend relevant lecture courses. Success in this demanding but preparatory ritual will be signified by the reception of a confirmatory vision.

Assuming that an encouraging vision was indeed received, the next stage began early on the first day of the new lunation. Four new leaves should be picked and the same solution prepared as before. Equally precise instructions are given for the recitations of prayers and the swallowing of the solution throughout this first day, and then for repeated recitations on each day of the rest of the month. Although the prayers should be said as often as possible, the practitioner may leave the house and engage in some aspects of ordinary life. A second month then follows, during which the procedure intensifies. During this stage, the practitioner must be at home at least from the third hour of the day. Moreover, special attention is given to the fourth, eighth, twelfth, sixteenth, twentieth, twenty-fourth, twenty-eighth and thirtieth days of the month. It is worth noting that this might place some limitation on the choice of the set of lunar months if it was to be literally observed. A month from new moon to new moon (a synodic lunation) has approximately 29.5 days, which led to the practice of alternating thirty-day (full) and twenty-nine-day (hollow) lunations in the Church's lunisolar calendar. However, additional lunations were inserted into the calendar at specified points in a cycle lasting nineteen years, and these added lunations always had thirty days. Thus, a sequence of two thirty-day lunations was possible. A simple solution to the problem would obviously be to take the last day of the lunation as equivalent to the thirtieth day.

Testing the commitment of the practitioner still further, a third month follows, during which the procedure intensifies yet more. Fasting and sleep deprivation continue, and sets of prayers are to be read or recited on twelve days of the month, with each set repeated up to fifteen times. Study of the appropriate subject should also continue. It is only in the fourth month that the procedure reaches its climax, with 'inspection' of the *notae* or figures. This requires seclusion with just one assistant or guide, and it may be assumed that the practitioner is by now in a highly receptive state. The creation and drinking of the solution containing the mysterious names, with accompanying prayers, is repeated as a prologue, and then, finally, come the figures of the chosen art. Once again the ritual must start early in the day, since the full set of figures (as many as seven for Philosophy) and all the accompanying sets of prayers, must be covered in one day. Moreover, all this must be repeated on each day of the month, with certain days being still more pressured, and with reading of portions of books dealing with the chosen art also still included. It is easy to understand how a practitioner might feel that nothing more sinister than a combination of hard work, spiritual observance and angelic guidance was involved in this very intensive procedure.

The angels are thought of in the text as directly involved in the process, and as having preferences as to when they will do most to help the practitioner. For instance, acquisition of the whole knowledge of the art of Grammar, some of whose figures have been illustrated above, will receive the most effective angelic help when the Sun is in Gemini. As Lat. 9336, folio 10r explains, 'the holy angels whose names are called upon have certain months and certain times in which they rejoice more than others'. This does not mean necessarily that the inspection of the figures for Grammar may only happen with the Sun in Gemini; but it does mean that appropriate angelic grace is more likely to be received then.

By the fourteenth century, this was not a straightforward matter, due to the phenomenon now known as the precession of the equinoxes. Bede's textbook on the construction and use of the Church's calendar, his *Reckoning of Time*, has a chapter on 'The Signs of the Twelve Months', which explains that each sign straddles two calendar months.[30] Thus, the Sun enters Gemini in May and leaves it in June. This makes the date of the transition an important issue. For Bede, the transitions took place in the middle of the month. However, Bede also states that the spring equinox took place on either 25 March (according to the Roman and liturgical calendars) or 21 March (according to skilled, pagan astronomers).[31] This was tricky since it was accepted from at least the Carolingian period that the Sun entered the spring sign of Aries at the spring equinox, and that each sign spanned 30° of the zodiac. For that reason, most illuminated calendars in manuscripts created up to the thirteenth century show the Sun changing sign around the twentieth of the month. But in the fourteenth century, the Sun was in fact moving from one sign to another much earlier each month, as the slight discrepancy between calendar year and solar year slowly accumulated over time.

For instance, the famous fifteenth-century Book of Hours, created for the duc de Berry and known as the *Très Riches Heures,* has the line marking the equinox and the beginning of Aries on the twelfth day of March. The image below also shows how dates of new moons and thus new lunar months could be calculated directly from this type of calendar. Crescent moons (indicating new moons) are shown on twenty days of this month and below each, in blue, is a letter between 'a' and 't'. The blue letters indicate the year in the Church's nineteen-year, luni-solar cycle in which a new moon will actually fall on that day of March (with 'a' signifying the first year in the cycle and 't' the last). Similarly, in this manuscript the calendar page for May shows the Sun entering Gemini on the thirteenth day of the month. For users of the *Ars notoria*, this ambiguity helped to support a practical shortcut, since the notes of guidance say that the whole of the month of May can be used for acquisition of the art of Grammar. And any month is possible, if the correct state has been attained.

The fact that the full procedure for attaining one art takes up so much of a calendar year would clearly limit the practitioner's rate of progress. Even if the introductory ritual were skipped, on the grounds that permission to proceed had previously been received, the fact that full days should be devoted – for months at a time – to the acquisition of one art would mean that several years were needed to complete just the seven liberal arts. This would be even more the case if the rules about new moons and zodiac signs were scrupulously observed, since suitable periods of four consecutive months did not occur regularly. Véronèse also points out that the text shows a considerable preference for the months from May to August as those in which the figures should be focused on.[32] This is practical, since the hours of daylight are longest in these months and the multiple repetitions of sets of prayers are thus most possible between the first and last hours of the day; but it would also add a further complication to the process of identifying a suitable period of time. It might even suggest that the favoured time for embarking upon a full ritual would be Lent, since that period of prayer and fasting would fit with the prescriptions outlined above, as well as being central to the lunar and solar calculations relating to the date of Easter.

Figure 6.8. Calendar miniature for March. Chantilly, Musée Condé, MS 65 (*Très Riches Heures*), f5v.

However, the practitioner who had achieved at least the preliminary vision and had become confident in pronouncing the necessary strings of strange words and names could make practical use of these achievements. Just as parts of the earlier procedures could be used as divinatory devices, so prayers from more advanced rituals could be used as tools to gain specified types of foreknowledge. Such enquiries tend to undermine the lofty assertions of spiritual purity found in the main rituals. They emphasise once again that the *Ars notoria* existed on two rather different levels. The more practical level, on which the *Ars* was a sort of toolkit for achieving a wide range of ends, included the possibility of getting answers to questions about life events or major decisions. An impoverished scholar could not only gain personal advantages in this way, but could also charge others for providing angelically inspired answers to their questions. The disadvantage would be that this would put the said scholar on the same level as professional magicians.

An example, discussed also by Véronèse, is found on folio 5r of Lat. 9336.[33] A short prayer is given here, the first part of which opens with an invocation of Christ and then of 'Hancor, Hanacor, Hamylos, Iehorna, Theodonos, Helyothos, Heliotheos, Phagor, Corphamados, Norizane, Corithico, Hanosae, Helse, Zope, Phagora'. The second part, introduced by a brief red rubric, adds another nineteen names and, on a less mystical level, requests the angels to tell the practitioner whether a given person will recover from, or die of, their illness. The marginal notes of guidance confirm the power of this prayer and call it 'the foundation of the whole art of Physic [Medicine] and of its mastery'. They go on to assert that, if it is recited in a suitably devout and humble fashion in the presence of other people, then certain types of truth will be revealed through the power of the angels. The main text accordingly sets out dialogues to be conducted with sick persons and procedures for dealing with pregnant women, or women whose virginity is in doubt. In all cases the truth regarding their likelihood of recovery, the outcome of their pregnancy or their possession of virginity will be revealed through the impartation of knowledge by the angels. If this sounds worrying, then the next section of the text reaffirms that this prayer was a gift from God to Solomon, to impart medical knowledge, and that it remains central to the art of Physic.

The aspiring student of magic would find wider-ranging guidance a little later in the text, under the heading of 'Astronomy' (on folio 8v in Lat. 9336). This term is taken to include astrology – and is explicitly stated to cover divinatory arts, such as geomancy and, more worryingly, necromancy. Explanations of each category of magical knowledge are given, including the information that necromancy involves the use of dead animals and their blood.[34] These magical arts are given considerable importance by the assertion that there are seven of them, just as there are seven liberal arts. Nevertheless, the facts that they are called 'adulterine' arts and that their place in the system of knowledge is somewhat uncertain make clear their difference from the mainstream arts. As in earlier manuscripts, they do not have complete *notae* of their own and are dependent on the 'general' images. The fact that the marginal text links them to Philosophy suggests an attempt to counter this; but it also suggests that Aquinas' fears were based on knowledge of such developments.

Evidence that the full version of the *Ars notoria* was attractive but over-demanding is found in the number of shortened and simplified versions that appeared in the later middle ages. These tended both to focus on more limited goals and to require considerably less time from their users. A popular promise was to improve memory, and some versions of the *Ars notoria* circulated under the title of *Art of Memory*. Fragments of one such survive in Kraków, Bib. Jagiellońska MS 2076. The tendency began in the fourteenth century, and MS Lat. 9336 itself has an introductory comment to the effect that, while Solomon called his revelation the *Ars notoria*, it is also known as the *Ars memorativa*. A different approach is shown in a manuscript given to St Augustine's Abbey by the monk Michael of Northgate. The manuscript itself has been lost, but its catalogue description includes an entry for 'Prayers taken from the *Ars notoria*', alongside a work attributed to St Augustine of Canterbury, a text on the angelic hierarchy, other prayers and guidance on confession. There is no mention here of the figures, and no suggestion that the *Ars notoria* was a suspect source for prayers. By contrast, the copy given by the monk John of London included '*Ars notoria* with Figures', and works on the magical rings of Solomon and on palmistry. However, even here the magical texts were in the unlikely company of the *Miracles of the Virgin Mary* and extracts from sermons and canon law. It certainly appears that the work was both widely available and open to different interpretations.

One of the most successful of the short versions was that known to modern editors as the *Ars brevis* (*Short Art*). Like the long, annotated version, this appeared in the fourteenth century; but it is noteworthy for its absence of any mention of Solomon or Apollonius.[35] It is also very different in its demands on the user, since its rituals require only a few days. In exchange for this, greatly strengthened powers of memory are promised, together with magical knowledge of future events, as in the full version. A surviving copy is found on folios 192r to 200r in the manuscript now British Library, Sloane MS 513, which was put together by Richard Dove, who became a monk at the Cistercian abbey of Buckfast in 1403.[36] Although the rituals here are short, each should still begin at a new moon and the practitioner should still confess, do penance, give alms, fast and pray devoutly.

Dove hoped to gain important advantages from this text, as the Latin heading to it shows. It gives the title *Ars notatoria* [sic] and promises understanding of all the arts, as well as 'the secrets of the whole world and the location of hidden treasure', to be revealed by an angel. Knowledge of future events is also promised to users, through a 'divine mystery'. Dove has enthusiastically added the exclamation 'Amen, amen, amen says Richard Dove, monk of Buckfast' on folio 192v.

Dove's text includes four simplified figures, offering the broad categories of 'Intellect', 'Knowledge', 'Memory' and 'Secrets'. One of these, on folio 199r, is (falsely) attributed to the scholar and theologian Albertus Magnus, whose work on natural philosophy and on the concept of natural magic had earned him a posthumous reputation as a magician. The ritual and *nota* for Memory begin on folio 195v. The figure given here is drawn simply in black ink and may be the work of Dove himself. It consists of two small, concentric circles, each containing a short piece of text taken from one of the prayers found in the full *Ars notoria*. They both have a plain, black cross at the top of the circle, marking their starting point, and the text of the prayer is given again outside the circles. The rituals have features in common with the full version, since they involve an early start and the repeated recitation of specified prayers. However, much more emphasis is placed here on the recitation of numerous masses, something that would be possible for Dove, since he was ordained a priest a few years after entering Buckfast. The devotion of the practitioner is further displayed by the requirement to recite prayers in one ritual with arms outstretched and supported on nails hammered into the wall, in imitation of the Crucifixion. Moreover, the preparation and drinking of the solution containing angel names does not appear here. Instead, the characters alpha and omega may be written on the practitioner's right palm before going to bed. That hand should be placed under the right ear and slept on, which should result in a revelatory dream. In other rituals, the *nota* may be copied out and then slept on, with related effects. It will be clear that all this has moved some way from the earlier versions of the *Ars notoria*, and yet it is still perceived as being somehow the same.

What can be said in conclusion is that this complex mix of liturgical rituals, ambivalent 'angel names' and challenging figures retained its fascination into the Renaissance and was apparently not affected by the Reformation. The text remained one of the most prominent works of ritual magic, and continued to be edited and abbreviated.[37] The promised rewards of both divine visions and intellectual triumphs clearly remained attractive, despite growing anxiety about the supposed activities of demonically manipulated witches. Part of the attraction may have been a hope of social mobility for those who could not afford to go to university. This is suggested by the fact that the *Ars* was translated out of the learned language of Latin, and also went into print. The early modern, edited version in English is still available.[38] The fact that printers felt safe in producing and selling copies of this work suggests that those who could afford and could use such a book did not feel vulnerable to accusations of witchcraft. That conclusion fits with the claims of practitioners of ritual magic that, far from being in the service of the Devil, they could issue orders to demons through the exercise of spiritual power. It also supports the view that a new, more educated and ambitious readership was growing for texts which offered increased knowledge and success through the use of magic. The rise of both amateur and professional practitioners, and their books, will be the subject of the next chapter, which will examine the increased visibility of ritual magic in the late middle ages.

7

KING SOLOMON AND
THE DEMONS

The Books of Professional Magicians

A growing number of magical texts, from the twelfth century onwards, claimed to embody the wisdom, and the magical powers, of the biblical figure of King Solomon. Many of these, like other texts considered in previous chapters, were translated from Arabic into Latin; but the growing demand for the magical rituals and procedures that they imparted is demonstrated by the rising number of vernacular versions. In many cases, the influence of – or direct borrowing from – Jewish magic is also clear. Several new texts, embodying ideas and procedures related to those in the older works, were claimed as the work of King Solomon.[1] This creates the impression that there was a recognisable category of magic that might be called 'Solomonic', after the king whose name was so often attached to it.[2] Care is needed, since other names were also associated with this type of magic, and the evidence is equivocal; but the term is useful in giving a name to a type of magical procedure that grew significantly in popularity in the later medieval and early modern periods.

At its heart was an emphasis on the summoning and control of spirits, angels and demons, an endeavour which pitted a human magician against beings of significantly greater power, and which therefore required thorough preparation and a range of magical tools. It is important to note here that this emphasis on what might be called 'active magic' could draw upon a long-standing tradition emphasising Solomon's powers over demons. It was also helpful that certain types of magical object, supposedly based on those owned and used by Solomon himself, were almost as important and powerful as written texts. This tradition is already strong in the work of the Jewish historian Josephus, in the first century CE. In Book Eight of his *Jewish Antiquities*, Josephus celebrated Solomon's mastery of the natures of all trees, plants and living creatures, and his ability (with the permission of God) to cast out demons. This knowledge was preserved in the large body of writings and incantations attributed to Solomon, some of which could bring about healing, according to Josephus. Others were used in exorcisms, including one which Josephus witnessed, in the presence of Emperor Vespasian. This exorcism involved the use of a ring – not to imprison a demon, but to cast it out. The ring gained its power through the combination of its inclusion of a seal and a plant root (based on Solomon's works) and the recitation of incantations composed by the king.[3] Objects linked to Solomon himself, or to the knowledge attributed to him, could be acquired by wealthy amateurs, as well as by professional users. They have almost certainly been lost in even greater numbers than copies of texts; but their importance may help to explain the tendency, which will become clear in this chapter, for 'Solomonic' texts to be treated as collections of tools and recitations for practical use, as much as learned treatises to be faithfully read and copied in their entirety.[4]

Such an impression is strengthened by the fact that from the middle of the thirteenth century onwards, philosophers and theologians who commented (usually negatively) on the growth of magical practice grouped a fairly coherent set of texts and titles under the name of Solomon. This group included Roger Bacon; the author/compiler of the *Speculum astronomiae*; and William of Auvergne, all of whom were polymathic philosophers associated with the expanding University of Paris. These 'Solomonic' texts are characterised by even greater variation in both titles and details than the other works of magic already discussed, but a recurring feature is their emphasis on the making and use of magical objects, such as rings and seals.[5] It is also the case that, despite earlier references, they survive only in (roughly copied) manuscripts from the later middle ages – and only very rarely in expensively produced volumes. Even treatises describing how to make relevant magical tools and images rarely include illustrations in medieval manuscripts. This suggests that the would-be magician was expected to be responsible for individual versions of the desired objects. This in turn implies that practitioners had the education, the means and the support required to devote considerable time to such activities. This chapter will therefore begin by examining surviving evidence and examples of Solomonic texts in what might be called 'professional' contexts. It will then move on to look at evidence from fifteenth-century and sixteenth-century volumes, which incorporate this type of magic into broader collections of medical, scientific and cosmetic materials. Surviving examples of the magical objects detailed in the texts are, unsurprisingly, even rarer; but evidence for their appearance and ownership will be discussed.

One of the earliest professional magicians to be clearly identifiable from both legal records and surviving magical texts is the Catalan or Valencian scholar Berengario Ganell.[6] He edited an extensive collection of magical texts under the title *Summa sacrae magicae* (roughly, *Encyclopaedia of Sacred Magic*) which was completed in c.1346.[7] The texts are not only edited, but also organised by Ganell, who includes comments on his own rituals and experiences. At one point he mentions carrying out a ritual while living in Perpignan, as well as showing knowledge of the University of Paris. Ganell appears in the documentary record as a magician linked to the court of James III, ruler of Mallorca, who was lord of Perpignan. Ganell's name was provided in a trial statement by Stephanus Pipini, who had once been a friar but was employed by James III as an alchemist. Pipini claimed that Ganell had been recommended as a 'master of these arts' and had given Pipini some training, as well as allowing him to make use of his books. If these were the texts on which the *Summa* was based, then access to such a library would be very helpful.[8] If it was indeed Ganell who had taught Pipini, then his magic had harmful, as well as more spiritual, applications: Pipini was accused of having made a wax figure of a bishop, with an intent to cause harm or death.

It is important to note that this accusation followed others that had been even more sensational, many of which had involved figures close to John XXII (pope 1316–34). Perhaps most famous is the case of Bartolomeo Cagnolato, who came to the papal palace in Avignon to level accusations against the Viscontis of Milan (who were engaged in a long-standing contest with the pope).[9] Two depositions were recorded, both in 1320, with the main accusation being that the Viscontis had questioned and tortured Cagnolato, believing him to be a magician, and had tried to force him to perform ritual magic with a silver image and poisonous suffumigations to bring about the death of John XXII. The case is best known for the fact that the poet Dante was also the subject of accusations. Only a few years earlier, in 1317, John XXII had also supposedly been the target of magical crimes, this time led by Hugues Géraud, the unpopular bishop of Cahors. In May 1317, Géraud was charged with participating in a plot to kill the pope and two cardinals, Gaucelin de Jean and Bertrand du Pouget, both members of John XXII's family. The first mention of such a plot appeared in a letter sent by John XXII to his nephews, Arnaud de Trian and Pierre de Via, on 8 March 1317. This letter recounted the arrest of two men from Toulouse at an inn in Avignon. They were found to be carrying suspicious packages containing tubes filled with white powder and three loaves of bread, in which had been hidden wax images of tonsured clerics. Each image was said to have been accompanied by a piece of parchment, on which were written the words, 'Papa Johannis moriatur et non alius, Bertrandus de Poyeto moriatur et non alius' and 'Gaucelmus Johannis moriatur et non alius' (i.e.

naming the three individuals who were to die). John XXII immediately ordered that investigations should be made and gave Pierre de Via and Arnaud de Trian authority to conduct the enquiries. Under questioning and torture, the arrested men named several employers and accomplices, leading to the accusations against Hugues Géraud.

The main witness told a complicated story, which included the confession that he was one of a group of four who had all gone to a Jew called Bernard Jourdain and asked him to make three images. One was a model of the pope, while the other two were dressed as cardinals, with hats that had been made separately. While they waited three days for the images to be made, they had gone to an apothecary called Durant Laurent, who lived near the chapel of Hugolin, taking with them toads, lizards, rats' tails and spiders; these they asked the apothecary to reduce to powder. They also presented him with a list of drugs they wanted. When they went back, the apothecary had put the drugs in several boxes and he gave them some other boxes with two or three empty compartments. He also gave them some goose feathers for handling the poison, and wrapped everything in two bags, which he closed round with sage and salt. The next day, two of the group – Prud'homme and Guillaume d'Aubin – had gone to 'La Saade', where executions took place, and cut the flesh from the leg of a dead man, along with some of his hair and nails and some of the rope used to hang him. They had also collected the tail of a dog they had found dead in the gutter as they returned to their lodging. It was this main witness who also provided the bulk of the testimony against Hugues Géraud.

At this stage Hugues, when questioned, only admitted to being involved in a plot to poison the pope. However, the sudden death of Cardinal Jacques de Via, another of the pope's nephews, led to further arrests and to another accusation against Hugues, who was now more harshly treated. '*Modicum tortus*' (moderately tortured) he confessed not only to having secret meetings in Avignon, but also to having received both an image of wax from Toulouse and the box of poison. He named a Jewish suspect; and he implicated the viscount of Bruniquel. On the evening of the same day he spoke of using images, one against Jacques de Via and the other against Bertrand du Pouget. However, his statements were becoming confused. Under further questioning, he confessed to bribing two papal knights to put poison, prepared by an apothecary in Montpellier, into the pope's food, to conspiring to receive an image, and to the pricking of the statue with the help of a Jew named Bonmacip. Hugues was condemned, degraded and then executed.[10]

Ganell's position was clearly very different from that of the unfortunate bishop of Cahors; but these cases make it clear that there was considerable evidence in the first half of the fourteenth century to support belief that serious practice of necromantic magic was taking place. Moreover, knowledge of magic, possession of magical texts and skill in magical rituals were apparently spreading among both the clergy and secular society. Ganell's book does provide considerable evidence of his status as a highly qualified scholar, studying and practising magic in a region not very far from Avignon. His introduction to the work asserted that magic required advanced knowledge of words and their uses, just like the more established subjects of Grammar, Rhetoric and Logic; and it also claimed that magic enabled the practitioner to bind spirits, both good and evil, through the use of the powerful names of God and of other entities.[11] His linguistic knowledge is suggested by the statement that magicians needed knowledge of Hebrew, Greek and Arabic, as well as Latin. However, that does not prove that he was the translator of all the texts included in his collection, since many were already well known. In addition to this linguistic prowess, the fully trained magician, according to Ganell, needed to be skilled in astronomy and astrology. Ganell also emerges as something of a historian of magic, since he saw a distinction between the 'old magic' of the pre-Christian period, of which Solomon was a leading practitioner, and the 'new magic' of Christians.

The *Summa sacrae magicae*, as would be expected, draws upon an impressive range of magical works. Its focus is on the summoning and control of 'spirits', which can be angels, demons or undifferentiated spirits. As it survives in a manuscript now in Kassel, it has five 'books' or major parts, each subdivided into chapters.[12] It opens with prayers, summonings and conjurations, together with crucial information on the preparation of the place in which the magic is to be performed and the circle which is to be used. These are followed by instruction on the powerful names to be used, especially of God, and of the ways to harness their power in relation to the winds and to locations on Earth. Times and seasons are important and are also dealt with in Book One. Book Two moves on to the equipment needed and includes the rings of Solomon; the crown of Solomon; the leash of Solomon; the sword of Solomon; the sigil or seal of Solomon; and the *candariis* of Solomon. This last term is of uncertain meaning, but it has been suggested that it denotes candles or candlesticks. Special 'figures' or images are also linked to the name of Solomon, as are chains. Books Three and Four deal with specific invocations, but the name of Solomon recurs in the listing for Book Five. Here Solomon is linked to the drawing-up and use of certain magical tables, which can be used in operations concerning the names of spirits. It is also in Book Five that the names of two other individuals seen as ancient authorities on magic are found, both frequently linked to Solomon. These are Toz Grecus and Raziel – the latter an angel said to have revealed a powerful work of magic to Adam. This text was claimed to have been subsequently edited by Solomon.

As a complete work, the *Summa* comprises a very thorough preparation and support for the performance of magical rituals, whose aims are made possible by the conjuring of supernatural beings. Despite the emphasis placed by Ganell on the special power of Hebrew characters and their incorporation into his tables of magical characters, recent analysis has suggested that he probably depended on Latin translations of Jewish magical texts.[13] However, it is certain that there are similarities between Ganell's descriptions of equipment for the summoning of spirits and those found in texts such as the *Clavicula Salomonis* (*Key of Solomon*).[14] The relationship is made complicated by the fact that surviving Latin copies of the Solomonic work are later than Ganell's work, but both the *Key of Solomon* and Ganell's version demonstrate the extent of the preparations and equipment required by a spirit-summoning magician. For a start, a minimum of seven special pieces of clothing was needed, each in the appropriate colour and formally linked to a specific planet. This is reminiscent of the works and practices of astral magic discussed earlier in Chapter 4. Two different types of crown are also detailed, made of different materials and bearing either actual stones or images of stones. The importance of the garments is emphasised by the fact that they are to carry names and seals, calling upon divine power. For those who could not afford a large collection of such robes, the names of God and of parts of the body of God could be written on parchment leaves, as needed, and attached to the inside of the garments in such a way that they would be in contact with the relevant parts of the magician's body. However, Ganell recommends that the divine names should be sewn into the robes in a more permanent way. After all this, it will not be surprising that the magician needs to be personally in a state of ritual purity and readiness, and that the place in which the magic is to be performed should be equally thoroughly planned and prepared.

Further evidence of the spread of professional magical knowledge and practice, and of the growing tendency for practitioners to produce compilations of their tested versions of established knowledge and rituals, comes from texts produced in southern France and in Italy. Important in this respect is the work of Antonio da Montolmo, who successfully pursued a career as a university master in late-fourteenth-century Italy. His best-known work on magic is *De occultis et manifestis* (*On Occult and Manifest Things*).[15] He also provided information on his own training in astrology and magic, recording that he had been taught by Tommaso da Pizzano, the father of Christine de Pizan. This is part of his account of Tommaso's use of astrological images to affect the outcome of a battle between the French and English armies.[16] Antonio also emphasises that such images are based on the work of Thebit ben Cora,[17] which was accepted by the author of the *Speculum astronomiae*, and therefore they are no more reprehensible than astrological medicine. The focus once again is on the summoning of spirits and angels, referred to by Antonio as 'Intelligences', which here are divided into four main categories, each linked to a specific sign of the zodiac. Timing and preparation are again crucial and are covered in detail, while a chapter is devoted to key pieces of equipment, namely images, rings and phylacteries (wearable boxes containing written names and words). These are made powerful by the performance of rituals of the sorts already discussed, but are most powerful of all when the Intelligences are successfully called upon to increase the strength of the astrological influences to which the images are linked. It is clear from this brief summary that Antonio's work takes an individual approach to his sources, including Solomonic texts, but his knowledge of these is obvious. For instance, he discusses the small, portable altar which he calls the *Almadel* of Solomon, and analyses the categories of angels that can be summoned with it.[18] Weill-Parot's analysis has found use of the *Four Rings* and *De angelica fictione* (both attributed to Solomon) and the *Idea* or *Key of Solomon*.[19]

Comparable magical erudition is demonstrated by a fifteenth-century volume now in the Vatican Library, Rome, which contains a selection of edited texts in Latin and Occitan.[20] Like Ganell's work, though (at least in its current form) shorter, this too was intended for practical use. Recent analysis has shown that its main treatise, the *Libre de puritats* or *Book of Purities* is based on the *Book of Raziel*, which was also used by Ganell. The term 'purities' here denotes the magical rituals expounded, and echoes the emphasis on the elevated state of mind required of a magician, as well as the elevated claims of the magic itself. The compiler of the Occitan volume remains anonymous, but seems to have been in possession of copies of several of the magical books translated and edited for Alfonso X of Castile, including the *Picatrix* and the *Book of Raziel* itself. As was customary, the central claim of this collection is that the carefully prepared and thoroughly equipped magician can both summon and control powerful, supernatural beings. Here the recitation of specified psalms is emphasised, while the use of appropriate suffumigations and body parts from sacrificed animals is more familiar.[21] The psalms are keyed not only to the seven planets, but also to the twelve signs of the zodiac and the twenty-eight lunar mansions. Thus, their recitation, at appropriate times, helps the magician to harness celestial forces. Knowledge of the particular powers of the various categories and ranks of spirits is also provided, again giving force to the promised effects of the rituals. For instance, the 'second army' of the first heaven can be summoned with the use of water and specific psalms, while the magician gains further power from the use of the complex divine name known as the Semiphoras,[22] which will be discussed further below. Once contacted, these angels can be called upon to bring about attacks on enemies and the sinking of enemy ships.[23] Still more impressive is the concept that angels may be called upon to put the magician in contact with the stars and thus, through resulting control of the winds, the magician can force demons to produce knowledge from the dead.[24] This is literal necromancy, in its original meaning of divination by means of the dead, yet in a form that emphasises the spiritual virtues of the magician.

Further examples of texts by late medieval experts on magic, who compiled their own edited versions of the art (or science) are being discovered and analysed by historians. One such is the work of a physician and astrologer from the early fifteenth century, Giorgio Anselmi da Parma, who gave his output the ambitious title of

Divinum opus de magia disciplina (*Divine Work on the Discipline of Magic*). The complete work survives in only one manuscript, written out in the sixteenth century by two scribes.[25] This volume is unlikely to be Anselmi's own work, which suggests that it had at least a limited circulation. Moreover, an extract survives in a later compilation, and it is possible that others remain to be identified.[26] Once again, the magician emerges as a person almost born to the vocation, who nevertheless requires thorough preparation, a secluded and specially built location for magical practice and equipment of types familiar from Solomonic texts. It is therefore not surprising that Solomon and his reputed works are cited at some length by Anselmi, even though he is selective in his attributions. Moreover, he sets out a view of the cosmos, its powers and the means of harnessing these that is essentially his own.

It is perhaps not surprising that specialists who studied and practised magic, often combining it with the provision of medical and astrological services to powerful clients, show both familiarity with Solomonic works and entrepreneurial adaptations of them. However, there is also evidence that, at least in early Renaissance Italy, some of this apparently specialist knowledge was making its way into manuscript volumes produced for wealthy, secular clients. These individuals were not themselves professional practitioners, and yet the excerpts, procedures and recipes incorporated into such 'books of secrets' appear to be intended for private use and are in the vernacular. One example has been linked to the court of Filippo Maria Visconti, ruler of Milan. The manuscript is dated to 1446 and is now Paris, Bibliothèque nationale de France, MS Ital. 1524.[27] Its editors have identified fifteen main texts, which appear to have been selected, translated and probably copied by one person. However, this translator-compiler does not seem to have been entirely in sympathy with the interests and choices of the patron who commissioned the manuscript. This is suggested by the notes and comments added in margins and even rubrics. Nevertheless, the production standard of both script and decorated initials is high, suggesting a wealthy patron. Moreover, it may be significant that, while the contents of the volume are drawn from a wide range of older treatises and works, their selection shows a clear emphasis on what is usually termed 'love magic'. Most of the rituals or 'experiments' in this category are anonymous here, although they are of familiar types. Evidence of the status of Solomon as an authority on magic is provided by the fact that a copy of the *Clavicula Salomonis* is included as the final main text.[28] Care was taken to establish a complete text, although the exemplar was difficult, and work by a second scribe is included here.

Given that surviving copies of the *Clavicula* are late medieval or early modern, this mention from 1446 of an old and difficult exemplar for the text may seem surprising. However, the work was known in Italy at least from the early fourteenth century, with the first witness to this being Pietro d'Abano. Pietro's *Lucidator dubitabilium astronomiae* (*Clarification of Astronomical Problems*) is dated to 1303–10, and in it he mentions familiar titles assigned to Solomon: the *Ydea* (or *Image*), the *Seals*, the *Four Rings*, the *Three Figures of Spirits* and the *Clavicula* itself. This is part of an impressively informed list of magical works.[29] At this point, if not before, it is clear that no one surviving manuscript can be taken as typical of Solomonic magic. Nevertheless, it is important to draw a distinction between the volumes centred on a limited number of relatively stable texts and the expanding format of the individually selected and heavily edited collection of necromantic materials. The latter became increasingly visible in the early modern period, in close connection to the fascinating figure of the Renaissance magician or magus, and both have received growing attention from historians.[30]

In these circumstances it is important to focus on a manuscript that can be fully viewed by readers who wish to see its complete contents for themselves; and so the volume that is now London,

British Library, Royal MS 17 A XLII will be central to the remainder of this chapter.[31] This is a handsome codex, presenting an English translation of an important medieval treatise of magic, written in a large and stately book hand, but without elaborate scribal decorations or use of precious metals in its illuminations. This volume is therefore also important for the evidence it provides of magical study and practice outside royal courts.[32] It has full-page illustrations depicting categories of angels, many of which are fully painted. The painting may perhaps have been done later, since, although it is valuable in depicting the colours associated with each type of angel, it is less skilfully executed than the ink drawings. Nevertheless, the fact that the text appears to have been specially translated and that these illustrations are unique makes this an important witness to new developments in the production of formal copies of magical books.

The volume opens in intriguing fashion, at least in its current state, since it has no title page and no title. Instead, it begins abruptly with an assertion that 'wicked spirits' have been plotting to 'send devils into the hearts of men' and have actually succeeded in corrupting the pope and the cardinals. The effect of this corruption is that the Church's hierarchy has outlawed the art of magic, here central to 'all things profitable for mankind'. Masters of the art of magic have responded by calling a great council of 811, seemingly mainly from the ancient cities of the Mediterranean, and have chosen one Honorius, son of Euclid, a master of Thebes, as saviour of the art of magic. Honorius in turn has been inspired by an angel, Hocroell (or Hocroel), and has assembled seven volumes of magical knowledge, divided into ninety-three chapters, 'in the which is briefly contained the effect of this art'. The resulting book is known as both 'Sacred' and 'Sworn', because it contains the sacred names of God, as well as the revelations of an angel; and because it can only be possessed by a master who has sworn to obey the stated rules about its use, protection and secrecy. In other words, this is an

English translation of the older Latin work usually known as the *Liber iuratus Honorii* (*Sworn Book of Honorius*).[33] However, a further twist in the story comes on folio 4v and links the book to Solomon. This is a statement by 'I, Honorius', in the name of Jesus Christ, that what follows comprises 'the works of Solomon', organised into chapters by Honorius.[34]

The dependence on Solomonic magic is immediately emphasised in the list of chapters which follows, since this states that the first chapter will deal with the great name of God, in its seventy-two-letter form. The 'Hebrew' title of 'sememphoras' is given. This name was an important element in the Seal of God, which is central to key texts and rituals of Solomonic magic; it is here described as 'the beginning in this art'. The remaining chapters of the first part of the text promise extraordinary spiritual powers for the successful magician, including not only a true vision of God, but also forgiveness of sins, protection from deadly sin and the ability to redeem three souls from purgatory. Possessors of this book, who presumably counted as members of the secret society of masters of the art of magic and were thus free from the demonic influences corrupting the church hierarchy, were clearly in an extremely privileged position. The doubts, turmoil and divisions of the advancing Reformation need not threaten their spiritual welfare. Such claims are startling, but the attribution of the original revelation to Solomon would be reassuring.

Part 2 offers an extensive repertory of practical information concerning the angels 'of every heaven', including their names, powers, seals, duties and rituals for summoning. This is followed by an impressive and enticing list of the types of all-encompassing knowledge to be gained from such angelic contacts. Part 3 comes a little closer to Earth, and focuses on the making and use of magical objects in order to exercise power over what are here called 'spirits'. These are mainly the familiar types of seals, tables (in the sense of devices for calculation, rather than items of furniture) and objects in which the spirits may be enclosed and controlled for future use. Once made, such objects can help the magician to control almost all aspects of the natural world and of earthly life, from weather and crops to the life and death of fellow humans, via control over animals and monsters. To a modern reader, the final chapters promised in this section appear rather less ambitious, since they focus on the power to conjure up illusions of traditionally magical and entertaining sorts, such as jugglers, musicians, castles, knights, gryphons and dragons. The final section is something of an afterthought, listing only five chapters and then retracting the last two because they are 'against the will of God'. Since these two supposedly instructed on summoning the dead and creating animals out of the soil, this is not entirely surprising (though it will be clear that this reservation would not apply solely to those two chapters). The list of promised (and withheld) chapters ends on folio 7r, and is followed by a heading in red stating that 'Here begins the Book.' This leads to an impressive and reassuring statement that the steps now to be followed are those of Solomon, and that the 'high mystery' to be achieved requires thorough spiritual preparation on the part of the magician, since the 'true invocation' is based on the names and powers and goodness of God.

There follows a declaration of the superiority of Christians as performers of magic – despite the importance of Hebrew letters and names and the dependence of the whole work on the knowledge of Solomon. The fundamental importance of the Seal of God is next set out in more detail than in the chapter lists, and is followed by the promised instructions on the actual making of the Seal. These instructions are copied out in Latin, with no attempt at an English translation. The implication appears to be that the translator regards these instructions as a highly technical matter, and as needing to be tackled directly by the would-be magician. It is certain that the instructions are given in great detail, since they occupy the lower part of folio 8v and then all of folios 9r to 13r. As would be expected from such a level of detail, this is a very complex assemblage of letters, signs and geometrical forms, all of whose placings and proportional relationships are important. In all, two circles, two crosses, a pentagram and three heptagons are to be assembled, in precise relationships to one another and to sets of characters. No diagram is given in the Royal manuscript, but an impressive version is included in the manuscript of Ganell's *Summa* (Figure 7.1).

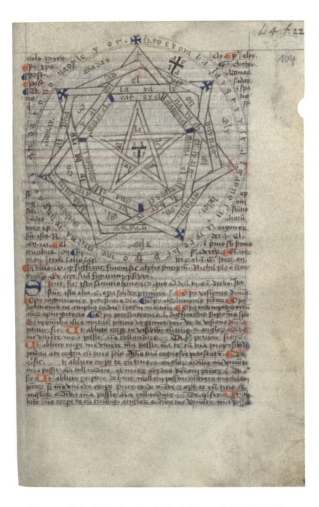

Figure 7.1. The Seal of God. Kassel, UB/LMB,
4° MS Astron. 3 (*Summa sacre magice*), f104r.

A sixteenth-century metal matrix, now in the History of Science Museum, Oxford, shows that demand for portable, three-dimensional versions of the Seal was sufficient to warrant the making of such equipment. The matrix is still in good condition and the museum also houses a plaster cast of the Seal (Figure 7.2).[35]

It is perhaps not surprising, for an object so concerned with the exercise of power over both celestial and material forms, that the colours and materials to be used for drawn copies of the Seal are specified in some detail. The text cites 'exemplars' as specifying different colours for different parts of the figure; for instance, the pentagon (or five-sided star) is to be red, with its inner space saffron-yellow. However, that seems to apply to a general or ideal version,

Figure 7.2. Seal matrix for the Seal of God.
Oxford, History of Science Museum.

since it goes on to say that for actual use ('in operations') varying supports and inks must be used. The 'inks' can be the blood of a mole, a turtledove, a lapwing or a bat (or a mixture), while the parchments can be made from the skin of a calf, a foal or a deer. It seems that the Seal was to be more solid than that suggests, since the instructions go on to say that the finished work is to be placed inside a special 'glove' and left outdoors on a calm night. From this it will be clear that the production of appropriate versions of the Seal involved a process demanding both skill and the ability to obtain quite specific components. It is also worth pointing out that the process of treating an animal skin to turn it into usable parchment took several days; and that of the parchments specified here, only that from a calf (usually known as vellum) was a standard commercial product. Accurate drawing of the geometrical forms would require at least a basic knowledge of geometry, as well as possession of the requisite tools.

As with the rituals found in the manuscripts produced for King Alfonso, discussed in Chapter 4 above, the 'incenses' to be burnt when summoning specified angels and spirits also required complex ingredients, adding to the demands on the practitioner. Both ritual purity and the wearing of special clothing are also important. The suffumigation for the Seal alone uses ambergris, musk, aloes, labdanum, mastic, ground pearls and frankincense. In effect, a magician would need a store of very expensive ingredients. Helpfully, and in the spirit of this type of magic, the editor of the Sloane manuscript adds in guidance on incenses and suffumigations from other textual sources.[36] A rule of thumb is that on Saturdays, incense based on strong-smelling plant roots are best; on Sundays resins and gums are needed; on Mondays leaves; on Tuesdays woods; on Wednesdays barks; on Thursdays fruits and seeds; and on Fridays flowers. When preparing incenses aimed at specific planets, blood and other body parts from appropriate birds and animals are also needed, though the resulting mixture should be firm enough to be shaped into pellets and should still smell strongly of the perfumes chosen. This guidance is attributed to a range of 'wise men' and philosophers, with Solomon playing a leading role. The same impression of wide reading and careful collation of material is given by the instructions for suffumigations to bring about visions and illusions, although at least some of these were less exotic in their requirements.

One recipe given on folio 17v promises to bring about 'secret visions' and revelations of coming 'hidden mysteries' affecting the whole world, and uses only flax seeds, violet roots and parsley roots, with a binding method to be revealed elsewhere. Rather more sinister is the recipe which starts on folio 20r. This demands aloes, whale sperm, costus (the more desirable form of this was based on Indian or Arabian trees, but an 'English costus' was also available), ammoniac gum, saffron, musk and the blood of a lapwing. As with all such preparations, an appropriate time for use needs to be calculated, taking into account the planetary day and hour and the zodiac positions of key planets. This would need to be done in advance, to allow for the careful preparation of the magician, as well as the suffumigation itself. The outcome promised is perhaps in keeping with the effort to be made, since 'visions in the air' should appear; and if tombs are suffumigated, then the visions will be of the dead.[37] It is reassuring to note that another recipe promises that a mixture of peonies, pennyroyal, mint and palma christi (probably the castor oil plant) will drive away visions and nightmares.

Further details on ingredients for suffumigations continue up to folio 22r, and are then followed by the promised next tier of information – that of the names of the angels whose intervention is to be called upon. The seven chief angels, each linked to a planet, its day and its celestial sphere, are listed as Capziel, Satquiel, Samael, Raphael, Amael, Michael and Gabriel. Many names have already been supplied in the details for the *Sigillum Dei* (Seal of God); inventive versions of Hebrew names are also given here. Much of this information draws heavily upon materials derived from Jewish magical works, and particularly upon those linked to the name of the angel Raziel. The key Latin works involved in the transmission of 'Books of Raziel' and the lists of angel names and terms of power which they made available to Christian magicians have a history at least as complex as those attributed to Solomon. In fact, the two streams of magical works converged, since Latin versions of the *Liber Razielis* frequently claimed that the knowledge, originally communicated to Adam, had been edited and transmitted by Solomon.[38] Organised lists of names, set out for specific time periods, including months of the Jewish calendar and days of the week, are given. With that in place, the focus moves on to the preparation of the blood to be used in creating the appropriate version of the Seal. This needs to be consecrated, and from folio 26r detailed instructions and prayers are given, citing Solomon as the source of this revelation.

The selections of prayers in this manuscript frequently use Latin, suggesting that the reader is expected to know some Latin, as well as to have access to help from a priest. This is the case in the set of prayers to the Virgin Mary, many of which would be familiar, as they are also found in Books of Hours. Prayers are given both for the consecration of the blood and for the preparation of the Seal itself, and the impression is that a magician would require a well-trained memory – perhaps especially for retaining the long and complex series of divine names. As editors have pointed out, there is considerable overlap between the sets of prayers found here and those provided not only in Books of Hours, but also in the *Ars notoria* and John of Morigny's heavily revised version of it (on this see Chapter 6 above). This wide-ranging use of relevant materials demonstrates that the compiler/editor of the Royal manuscript was an expert in his subject, as well as sharing the view that texts of angel-summoning magic could be treated as repositories of materials to be reused at will. The suggestion that this manuscript represents a deliberate creation, aimed at use by a practising magician, is further supported by the instructions given in rubrics between groups of prayers. These acknowledge that the primary aim of much of the material (especially that derived from the *Ars notoria*) was to achieve true vision of God – but they also advise on switching to an aim of the magician's own choice. Many of the materials given are long and complex, as was also the case in the *Ars notoria* itself; but the overall impression of this manuscript is of greater practicality in approach. Specific instructions are given, on folios 29r and 34r, for practitioners who wish either for secret or privileged knowledge or to consecrate their book. More elaborate directives are given for those who wish to summon a spirit. Similarly, instructions in English are given on folio 44r for sequences of prayers, culminating in a series running from Friday through Saturday night, at the end of which the magician will be told by an angel whether he (it is always 'he') can proceed to the next stage.

For those able to move to the next stage, more detailed goals are covered. As well as those already outlined, these include knowledge; mastery of arts; protection from evil; knowledge of the heavens; and knowledge of the angels and their seals. At this more advanced stage a priest is also needed, and it is interesting that the text seems to assume that the magician will not be a priest – or at least will not be able to perform the necessary services, while also continuing with the rituals. Further long series of prayers, focused particularly on powerful names of God, are given up to folio 58v, where attention shifts entirely to the attainment of the vision of God. This section ends on folio 65r, with an editorial comment that guidance for attaining the divine vision is now complete. However, those who wish for knowledge of divine power, or freedom from sin and its consequences, or to redeem three souls from purgatory, should follow the instructions about to be given, in order to adapt the text to their requirements.

A break in the text follows, marked by a somewhat confusing heading, which states that what comes next is the 'Second Treatise of the Second Book'. As might be expected, the chapter list given next accords with what was promised for Book Two at the start of the manuscript. However, the heading at the bottom of folio 66r states: 'Here follows the treatise of the nature and office of the second angels.' These angels, it seems, are distinct from the orders of angels whose affiliations and names have already been expounded. Moreover, they love and serve God above all else (as is only to be expected), but can be sent to chosen humans. As might also be expected, there are seven main groups of these angels, each linked to a specific planet and each with a distinct nature and office. Details of each group are given, starting with those of Saturn, the outermost planet, and moving through the Ptolemaic order to end with those of the Moon (on folio 69r). An oddity of the lists is that several names, such as Satquiel, Raphael and Michael, are repeated. What is unique to this manuscript is not the listing of names or the characters or seals linked to each, but rather the provision of fully painted images of the leaders of each angelic group. Folio 67v depicts both the angrily red angels of Mars and the gold-glowing angels of the Sun. The Seal of the angels of Mars can be seen, although that of the angels of the Sun follows on the next folio. It is noteworthy that these beings are here called both angels and spirits – but never demons.

Armed with a full repertory of ritual instructions, prayers and invocations, angel names, seals and suffumigations, as well as permission to proceed and the services of a priest, the magician is now ready to construct the necessary magic circle. Guidance on this begins on folio 69r and is further evidence of the level of commitment and expenditure required for this type of magic. The outer circle must be constructed with shaped and cleaned stones or special tiles, joined with mortar using clean sand from the seashore. The circle itself must be level and must be 14 feet in diameter. Inside it must be a smaller, broken circle, 7 feet in diameter and 3 foot 6 inches high. The expense of this is acknowledged, since the text allows that magicians may, if needed,

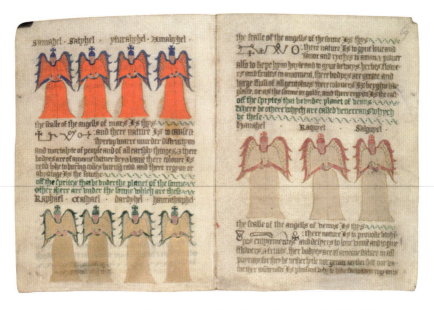

Figure 7.3. Angels of Mars and of the Sun. London, British Library,
Royal MS 17 A XLII, f67v.

substitute an earthwork of the same dimensions. Once complete, the
circle – and the magician – must undergo a three-day cycle of cleansing,
at the end of which the magician should be completely shaved, before
donning white clothing. He should hold the Seal in his right hand.
This must be carefully timed for the appropriate day and hour, in
accordance with the angel or spirit to be summoned. Similarly, the
magician must face in the correct direction. The support of the priest
is important since it is he who prays for God's protection for the magician during the summoning that is to follow.

A space of considerable size is needed for the summoning itself,
since the magician is to draw two smaller circles in the ground with
a clean knife, and these should be a foot apart. Inside them are to be
written 'the names of the angels of the hour, of the day, of the month,
of the time, and of the face' (the division of a zodiac sign). When
the invocation itself begins on the third day, the magician should be
standing outside the circle. This may suggest that the circle is the

'seat of Samaym' at which the angels are to appear. The role of the smaller circles is not entirely clear. The relatively practical approach of the editor/compiler of the Royal manuscript emerges again at this point, since, as Joseph Peterson and other editors have observed, it now departs from the longer text of the *Sworn Book of Honorius*. Presumably feeling that sufficient material has been drawn from that source, the text on folio 71r moves on to the somewhat vague heading 'Another work for [summoning] the angels of the seven heavens, and the four seasons of the year'. This section is also illustrated with images of the angels, each named. The contents go considerably beyond what the heading promises, since the angels of the heavens and the seasons end on folio 73r, and are followed by angels of the twelve signs, of the twenty-eight mansions of the Moon, of the four winds and of the four elements. All these are illustrated with drawings of familiar types of angel.

The next section, listing seventy-two angels and referring to the 'Schemhamphoras' (here spelt in a more usual form than earlier), is not illustrated. However, research among sources is again suggested since reference is made to 'the Cabalists' and to 'Tables of Ziruph'. This precedes an apparently unfinished additional section on angels, starting on folio 76r and concluding on folio 78r. The folios are filled with outlines of angels, but only the first thirteen of these are named. A further suggestion that this part is unfinished is that an empty space is left on folio 78v. The impression that this last section of the volume is not in its final state is increased by the last folios, 79r to 80r. These provide further angel names, attributed to 'Zevell', additional 'Princes of the Twelve Altitudes', and then additional sets of angels for the planets and the signs – attributed to Honorius – as well as more for the lunar mansions, the winds, the elements and the Schemhamphoras. There is no attempt at a conclusion; instead, the simple word 'FINIS' is found on folio 80r.

It should be noted that the sources for this supplementary section are even more complex than is suggested by modern editors. The latter suggest that the borrowings were mainly taken from the recently published, printed version of the multi-volume work *On Occult Philosophy* by Heinrich Cornelius Agrippa (including a later, fourth volume spuriously attributed to him) and a similar survey of magic equally spuriously attributed to Pietro d'Abano.[39] However, the situation is, if anything, even more complicated than this suggests, since those works themselves necessarily drew on older works, including the *Sworn Book of Honorius* and the *Ars notoria*, which are themselves inter-related and which are also directly used by the editor/compiler of the Royal manuscript. At this point an exact identification of the immediate source for each part of the text becomes impossible. But fortunately it is not needed. The most important finding for the current study is the care taken in the Royal volume to provide a full range of materials, textual and practical, for the enthusiastic practitioner of Solomonic or angelic magic.

Also significant, and suggestive of this work as marking a transitional point in the history of such magic, is the dominance of English in much of the text. The continuing use of familiar Latin in prayers shows a traditional respect for those prayers and the efficacy attributed to them, matching the use of terms like 'Mass' and the belief in Purgatory. Equally conservative is the absence of dependence on crystals or reflective surfaces to support the apparition of the angels. That technique is recorded in descriptions of magical practice from at least the twelfth century, but became dominant in ritual magic and in the summoning of spirits only at the very end of the medieval period. Yet the use of the vernacular and the slight clumsiness in both the script and the illuminations of this manuscript suggest a rather new type of owner for the volume. This does not seem to be a book for the professional scholar and 'philosopher-practitioner' on the one hand, nor for the princely connoisseur of esoteric volumes on the other. Instead, an image grows of an owner from a slightly less elevated section of society, who nevertheless wished to benefit from direct involvement with magic and to gain the tempting rewards promised in the opening folios of the volume.

What is particularly striking about this, in relation to the broader history of magic, is the evidence it provides that the perceived efficacy and value of magic outweighed the dangers (both spiritual and legal) of its practice. As Klaassen has argued, early modern ritual magic drew heavily on the magic of the medieval period, even as its social base widened.[40] An equally important point is that, while this type of magic was subject to penalties, it remained distinct from the category of witchcraft, whose rise and dire consequences have been discussed in detail by historians. For this book, what is striking is that neither fear of witchcraft nor reaction against the spread of magic stopped patronage of magic among the elite. Royal courts in fifteenth-century Europe were as likely to house volumes and practitioners of magic and predictive astrology as were those of the thirteenth and fourteenth centuries. The final chapters of this book will examine this phenomenon.

8

CHRISTINE DE PIZAN AND ASTROLOGY IN THE FRENCH COURT

The previous chapters have traced the increasing availability of complex Latin texts on ritual and astral magic in learned circles. It has also become clear that considerable numbers of professional and would-be-professional practitioners of magic, predictive astrology and prognostication were present in centres of power, including even the papal court. Under such circumstances, it is not surprising that there is evidence of the employment of such individuals. The desire for knowledge concerning coming developments and dangers, which has been a thread running through most of the chapters of this book, was still strongly present, despite rising warnings from moralists and theologians. The range, complexity and perceived usefulness of astrology put it at the heart of these competing developments. 'Modern' medicine and natural philosophy made heavy use of astrological, as well as astronomical, procedures and theories. But an expert in these valued fields would also be capable of producing personal predictions of more problematic sorts. The resulting tensions have been studied by historians for some time, but less attention has been paid to the question of gender and the extent to which women had access to knowledge of astrology, in particular.[1]

The exceptional career of Christine de Pizan as a writer, translator, scholar and adviser to the powerful has been increasingly and deservedly celebrated. However, the main focus has understandably been on her courageous and powerful arguments for greater recognition of the virtues and achievements of women. The energy and courage which she also displayed in standing up for the value of astrology have been less noticed. There is no suggestion that Christine herself was a practitioner of horoscopic astrology, but it is striking that her writings display considerable expertise in astronomy and astrology. Moreover, she was very positive about both the social value of astrology more broadly and the personal interest displayed by King Charles V of France. The fact that Christine's father, Tommaso de Pizzano, was a master of medicine and astrology at the University of Bologna and was personally invited by Charles V of France to join his household has received attention, as has Christine's strong admiration for her father.[2] Her skill in praising the study of the stars, while avoiding scandal when illicit astrology was attacked by leading churchmen, has also been noted.[3] Nevertheless, a reading of her works, and the manuscripts which she presented to powerful rulers and patrons, produces fresh evidence concerning her own knowledge of astrology and the means by which she evaded accusations of illicit – or even demonic – knowledge.

Christine's case for astronomy and astrology is carefully set out in her biography of Charles V, and is part of her praise of him as the 'wise king'. The work was commissioned and written during the wars and troubles of Charles VI's reign, and was completed in 1404.[4] The third part is entirely devoted to recounting evidence of Charles' wisdom and knowledge, drawing not only on the works of the paramount philosopher, Aristotle, but also on Thomas Aquinas' *Commentary on Aristotle's Metaphysics*.[5] Almost from the start, in Chapter 4, the king is celebrated as an expert in the science of the stars. The chapter is headed: 'King Charles as an astrologer and the nature of astrology'. Christine goes on to assert that Charles was 'very expert and learned in advanced philosophical subjects like astrology' and that he 'understood clearly all the key points of that subject and loved it as something special'. In the same chapter, she shows personal knowledge of the scientific and mythological works of Ptolemy of Alexandria, Plato, Ovid and more recent authorities, such as Avicenna. This level of formal learning on the part of a woman is truly exceptional – and its public display even more so. The sources are used, also boldly, to expound the case for the priority of the study of the heavenly bodies as the best route to knowledge of the First Cause of all celestial movement – God. Even more impressively, Christine made use of what appear to be her own translations of these Latin texts, although she was too modest and too cautious to say so.[6] It goes without saying that she took equal care not to argue for problematic applications of astrology.

Interestingly, Christine does suggest differences in the king's level of expertise in different branches of the seven liberal arts. Charles emerges as naturally gifted in rhetoric, logic and geometry, while being attracted to, and carefully taught in, 'astrology'. That subject is here defined as sure knowledge of the movements of the celestial spheres and of the planets.[7] When recording the translations of major works that the king commissioned from leading scholars, Christine emphasises the books of the Bible, with glosses and theological commentaries, together with works of theology, as would be expected. Equally understandably, the works of Aristotle immediately follow. However, Charles' own scholarship and his building of a great library are compared to those of 'Ptolemy Philadelphus' of Egypt.[8] This figure was a compound of a historical ruler of Egypt and the astronomer/astrologer Ptolemy of Alexandria, a confusion widespread in late medieval chronicles. The effect here is to emphasise again the strong link between wisdom, book collection and knowledge of the movements of the heavens.

Christine also recorded, in Chapter 22, that the king was deeply interested in alchemy – and especially in the work of a master from Avignon who was reputed to have discovered the philosopher's stone. However, it is astrology that receives the most attention in her account. Charles' achievement in staging a formal meeting with the emperor in Paris, and in holding discussions, is recounted in several chapters towards the end of the book. These include details of the costly and impressive gifts given and received; and the list of gifts handed over by the French side is headed by a great, gold cup presented by Jean, duc de Berry. What made it especially precious was its decoration in enamel-work, depicting the zodiac, its signs, the planets, the stars and the constellations.[9] Finally, the book concludes with a series of chapters on the subject of the knowledge, virtues and wisdom needed by kings and from which their kingdoms can benefit. Chapter 67 returns to the argument, again citing Aristotle, that study of the heavens and the celestial bodies was fundamental to the rise of philosophy and human knowledge. Therefore, for Christine, astrology (in this broad definition) has priority over all other branches of knowledge.

Christine's outline of astrology in this work is effectively that of astronomy, echoing widespread medieval usage. However, elsewhere she asserts that skilled astrologers could gain knowledge of developments on Earth, as well as glimpses of the power of God, from study of the stars. Indeed, she argues that it is the duty of rulers to pay attention to the advice that such astrologers can give – although the difficulty of interpretation is such that rulers should employ several astrologers, not just one. For instance, in *The Book of the Body Politic* (written 1404–07), Christine once again uses Aristotle and Ptolemy, as well as ancient experts like Archimedes, to support her belief that good rulers should include astrologers in their privy councils. Frauds are obviously to be avoided, but since it is universally acknowledged that 'this earthly world below' is subject to the influences of the heavenly bodies, it follows that accurate analysis of the stars can be 'very profitable in the maintenance and improvement of cities and kingdoms'.[10] A strong note of caution sounds when Christine goes on to emphasise that no prince should consult anyone who uses 'evil arts forbidden by the Church'; but what she calls 'pure' astrology is entirely natural in subject and process and uses neither evil arts nor any sort of incantation.[11] The latter was crucial, since theologians from Aquinas and Bonaventure on had developed the concept of the unwitting pact with a demon, in which uttering ritual words played an important part. It is significant that in medieval Latin the concepts of incantation and enchantment were closely related. For Christine, there is no suggestion that a good astrologer would have contact with dangerous forces and spirits, as users of the *Ars Notoria* and *Picatrix* did.

Such openly magical texts may appear far from the expected contents of a royal library, especially one open to members of the court, royal counsellors and distinguished visitors, as the library of Charles V was – but that is contradicted by the evidence of the catalogue of Charles' books. This was drawn up in 1373 and is notable for paying as much attention to the richness of the jewelled and decorated bindings of many books as to their textual contents. The compiler, Gilles Malet, seems to have been a knowledgeable scribe himself, since he gives details of the type of script used in many volumes; but his information on pictures and illuminations is sparse. The catalogue makes it clear that by 1373 the king's books were well organised in a three-storey tower in the Louvre, which also contained a luxurious study for the king's personal use. Books recently acquired, and especially newly translated works in French, were housed in the first chamber. They included a set of volumes, many in matching bindings of rich silk and with elaborate silver clasps, entered as numbers 57 to 62. These were shelved with works on law and on history, and all contained texts of astrology and geomancy.[12]

Ptolemy of Alexandria is represented by his astrological textbook, the *Quadripartitum*, as well as by the *Centiloquium*, which it was mistakenly believed was by him. The multi-author work titled the *Book of Nine Judges* was present, as were treatises by 'Haly Abenragel', 'Zael' and 'Messahala'. The first of these is Abū al-Hassan 'Alī ibn Abī al-Rijal, and his work was probably the *Book on the Judgements of the Stars*. The second was Sahl ibn Bishr al-Isrā'īlī, a Jewish astrologer and mathematician, whose works also provided wide-ranging guides to judicial astrology. Messahala was the medieval Latin name given to Māshā'allāh, many of whose works were available in Latin.[13] Such a collection clearly included the problematic type of astrology known as 'judicial astrology', which had been criticised by theologians from Augustine onwards. It was problematic not only for contradicting key Christian teachings on free will, but also for opening practitioners to the dangers of demonic temptation. It might be suggested that this set of books was given to Charles as a gift, rather than representing his own tastes and interests; but evidence contradicts that.

It was standard practice by the late fourteenth century for elite individuals to employ physicians who were skilled in astrology, since the two disciplines were closely inter-related and astrological medicine was entirely acceptable. Christine de Pizan's father, Tommaso, was personally recruited by Charles V to safeguard his health in this capacity. However, the skills needed by such highly trained physicians would also enable them to provide confidential answers to questions of more controversial types. Charles V, both while dauphin and as king of France, openly employed astrologers: as early as 1361, he commissioned from the 'astronomien' Robert Godefroy, a French version of the astrological textbook *The Book of Nine Judges*, and it is likely that the volume entered in the king's book catalogue at number 59 was a luxury copy of this work. Another astrologer employed by Charles, Pèlerin de Prusse, produced both an astrological treatise and a guide to the astrolabe, finishing that work in 1362. His prefaces express due modesty, but editors of his works note his familiarity with classic tracts on astrology, his confidence in the casting of horoscopes and his pride in his ability to offer novel techniques and procedures. His popularity with his royal employer is clear from the facts that he is called 'beloved clerk' in the dauphin's records, and that accommodation for both Pèlerin and his servant was provided in the Hôtel St-Pol in Paris. Pèlerin is referred to as 'master', which suggests recognition of his skill and qualifications.[14] The recruitment of Christine de Pizan's father provides further evidence of Charles' high level of interest in such skilled practitioners.

Still more striking is the further evidence provided by analysis of the full contents of the catalogue of Charles' library.[15] An attempt will be made here to distinguish between works that would now be considered astrological and those that were clearly astronomical, despite the ambiguity of these terms in the fourteenth century. The very fact of Christine's carefully broad terminology makes this an interesting issue. As would be expected, the astronomical works entirely bear out the claim that Charles was well instructed in astronomy. Ptolemy's fundamental and very technical work on planetary motions, the *Almagest*, was present in no fewer than four versions, in both Latin and French, and even the *Commentary of Geber* was included. The latter work was translated into Latin by Gerard of Cremona (who also translated one version of the *Almagest* itself).[16] It was very much a work for specialists, since it drew on revised geometrical models to correct Ptolemy's original.

Astronomy is represented in a total of thirty-two volumes, not including the *Almagests* already listed. This total includes Nicole Oresme's French version (incorporating commentary) of Aristotle's *On the Heavens*, as well as copies of the major textbook, Sacrobosco's *The Sphere*. Works on the astrolabe are listed as present in three volumes, as are three versions of the *Theorica planetarum* – a title given to competing textbooks on more advanced astronomy.[17] Where treatises are simply described as dealing with 'the zodiac', it is likely that they provided basic information on both astronomy and astrology, but they have been counted as works on astronomy here. Beyond Ptolemy and the compilers of Latin textbooks, the author most frequently named is Alfraganus (found in three volumes: 619, 830, 831). This is the medieval Latin form of the name of Abū al-ʿAbbās Aḥmad ibn Muḥammad ibn Kathīr al-Farghānī, a ninth-century astronomer best known in the Latin world for his revision and summary of the *Almagest*.

Whether these volumes were used by the king himself may be doubtful, and the same applies to the numerous copies of sets of astronomical tables. The Alfonsine Tables are specified in five volumes, with one set of English Tables and five anonymous sets of tables adding to the total. There are also some 'canons' or rules for using the tables and a separate set of eclipse tables. Ready-made applications of such tools are also found, in the form of two copies of the *Almanac* of John of Saxony, a fourteenth-century master at the University of Paris.[18] There was also one anonymous almanac. These tables were extremely useful, since they gave planetary positions for each planet, on a daily basis if appropriate, and relevant for a given location. John of Saxony's original work ended in 1380, but was extended in some surviving manuscripts. Overall, it is clear that many volumes would have been primarily of use to the court astronomers and astrologers, but that more basic instruction in the subject, and its applications, was also perfectly possible within the royal household. The same goes for mathematics and geometry, since eight volumes, two in French, can be placed under these headings.

Still more impressive is the number of astrological works recorded in the king's library. Identification, where neither author nor specific title is given, is subject to the same limitations as for works of astronomy; but something like fifty-six volumes can be placed entirely or mainly under this classification. As might be expected, Ptolemy's name occurs most frequently, with nine entries, excluding the *Almagest*. Several of these related to copies of the wide-ranging 'guide' to practical astrology, wrongly attributed to Ptolemy and known as the *Centiloquium*. This, too, was translated into Latin in the twelfth century and was one of the most popular textbooks on its subject, covering both more general (or 'mundane') astrological readings affecting large numbers of people and the issues most frequently affecting individual clients.[19] It was also useful in providing techniques for dealing with problems caused when clients were unable to provide important information, such as the date and time of their birth.

Next most popular in Charles' library was 'Albumazar', with seven entries, while Alcabitius, Messahala and Haly Abenragel had six entries each. Alcabitius' *Liber introductorius* (*Introduction to Astrology*) was the main textbook used by the late fourteenth century in universities, including Paris. Charles' collection clearly included astrological knowledge going beyond what was taught to most students.[20] The importance of astrology for medicine is clear, since twenty volumes on medicine were also included, many focusing on medical applications of astrology. It is difficult to be more specific, since several were listed simply as 'On Medicine'. Returning to astrology, it is worth noting that the royal library included not only most of the 'classics' on the subject, but also more recent works, which offered tips and new material from personal experience. Books in this category included Guido Bonatti's work (probably the *Liber astronomiae* or *Book of Astronomy-Astrology*) written in c.1277. Bonatti was famous (or perhaps infamous) both for his successful career as an adviser to rulers in Italian states and for his condemnation in Dante's influential *Divine Comedy*. Comparable and equally controversial would be the work that circulated under the name of 'Leopold of Austria' (c.1271). This was grandly titled *Compilation of the Science of the Stars* and boasted of covering the most important parts of astrology in its ten books. The actual identity of Leopold is not certain, and the book itself says only that he is a 'son of Austria'. Its preface boasts of the long study in which the author/editor has engaged, as well as of the powers which God gave to the celestial bodies and to which it offers access.[21]

However, 'mundane' astrology, especially works relating to the forecasting of weather and natural disasters, is also well represented. The value of paying attention to warnings concerning such matters had biblical support in stories such as Joseph's interpretation of Pharaoh's dreams, as recounted in the book of Genesis, even if King Saul was severely punished for consulting a woman who had a 'familiar' spirit.[22] Concern as to the distinction between acceptable and unacceptable forms of astrology is represented by the presence of a French translation of the *Speculum astronomiae* attributed to the theologian and scientist, Albertus Magnus.[23] This thorough scrutiny of versions and texts of astronomy, astrology and astrological magic was produced in the second half of the thirteenth century. It probably informed the formal condemnation in 1277, at the University of Paris, of a list of forbidden works and beliefs. Unsurprisingly, it condemned forms of astrological magic involving incantations, mysterious names and engraved characters. Nevertheless, it supported the view that the stars and planets possessed powers that were part of God's creation. For that reason mundane astrology – including astrological medicine – was entirely acceptable. Nevertheless, it is clear that Charles V owned many astrological texts that would be condemned by the *Speculum.*

This conclusion is strengthened by the fact that the king's library contained numerous works on astrological magic, and even a volume on necromancy, as well as less controversial texts on palmistry and collections of prognostics. Alchemy was represented by only two complete volumes, while palmistry accounted for four (with one in French), and appeared in three further, compound volumes. The collection of prognostics sounds rather like that put together by Matthew Paris of St Albans in the previous century.[24] Three volumes contained predictions or 'judgements' attributed to Socrates and Pythagoras, among others, sometimes accompanied by other forms of divination. Much more popular than these, and representing an increasingly fashionable form of divination, were works concerning geomancy. This technique, associated with astrology in many versions, could be found in no fewer than nineteen volumes, mainly as the only named text. One is specified as being in French and another is attributed to Jean des Murs, but the majority are distinguished only by their scripts and bindings.[25]

Another five books are openly or probably magical, with the French *Necromancy* appearing at 741. No further details are given.

There are two copies of the *Ars Notoria* (at 751 and 908). Entry 595 describes a volume of texts in French, one specified as the work 'On Images' by 'Ben Thebit ben Corab',[26] another dealing with the Rings of Solomon and yet another being a mysterious 'Psalter of Constantinople'.[27] The title given to the work by Thābit ibn Qurra suggests that this is based on the Latin translation of his treatise on astrological images, made by John of Seville in the twelfth century. This version was regarded by the compiler of the *Speculum astronomiae* as acceptable, since it did not include potentially demonic elements. However, a slightly earlier translation by Adelard of Bath, which was given a title usually translated into English as *Book of Talismans*, was condemned.[28] Numerous treatises on magical rings, several of them attributed to Solomon or linked to his name, circulated in the later middle ages. The most popular seems to have been one simply entitled 'On the Rings of Solomon', which gave instructions for making rings under the powers of each of the twenty-eight mansions of the Moon in turn. This may well have been the work included here; although there was also an apparently well-known treatise on the 'Four Rings of Solomon', which dealt with rings to summon spirits and demons.[29] A French version of the *Book of Secrets* – believed to be by Albertus Magnus and dealing with the marvellous powers attributed to herbs, stones, beasts and planets – was also included, along with 'many other things'.[30] As tantalising as the French *Necromancy* is volume 715, simply entered as *Mathesis*. This term was discussed by the twelfth-century bishop of Chartres, John of Salisbury, in his *Policraticus*, a work on good government, which was translated into French for Charles. According to the learned bishop, the word could denote either acceptable mathematical knowledge or its unacceptable, astrological form, depending on how it was pronounced. It may be relevant that works on mathematics are usually classified in this catalogue as 'algorismus'.

Comparison of Charles' library with that of his brother-in-law, Gian Galeazzo Visconti of Milan, confirms that Charles' level of expertise in astrology and magic was unusual, even at a time when such interests were fashionable. Cultural exchange between the two courts was almost certainly supported by inter-marriage. Charles' sister, Isabella of France, married Gian Galeazzo Visconti in 1360, a week after her twelfth birthday, and may have been the donor of French translations of authoritative works. Isabella and Gian Galeazzo's daughter, Valentina, married Louis, duke of Orléans in 1388, and thus became part of the rival factions vying for power in France. Valentina brought twelve books with her, almost entirely service books and works of piety, but left a collection of forty-three volumes when she died in 1408.[31] The Visconti library was catalogued in 1426, when it held 988 titles.[32] It, too, contained two versions of the *Ars notoria* (one attributed to Solomon), as well as two volumes on alchemy. However, there are no works of 'necromancy', no openly magical treatises on rings or related subjects and far less interest in forms of divination. Geomancy appears only twice and chiromancy not at all (though it could be concealed within collections on natural philosophy and 'secrets'). Interest in astronomy and astrology certainly appears; but there is only one copy of the *Almagest*, one *Centiloquium* and one copy each of the Tables of Toledo and Alfonsine Tables. The number of volumes clearly (or probably) on these subjects is only forty-four, a total far exceeded by those on medicine.

Other evidence, however, suggests that Gian Galeazzo was no less likely to consult astrologers than Charles: simply he was less interested in studying the subject himself. Philippe de Mézières, counsellor of Charles V, tutor of Charles VI and ally of the duke of Orléans, celebrated in his *Songe du Vieil Pèlerin* the extraordinary achievements of the astronomer, astrologer and horologist, Giovanni de Dondi of Padua.[33] Inspired by astronomical study and a desire to make celestial and planetary movements on any given date visible almost at a glance, Giovanni constructed the technological marvel known as the *astrarium*. Philippe gives an awestruck account of this mechanism, which took sixteen years to build, and adds that Giovanni and his astrological machines were in great demand in courts in Italy, Germany and Hungary. Gian Galeazzo Visconti was known to pay him an impressive salary amounting to 2,000 florins per year. Valentina Visconti, Gian Galeazzo's daughter and duchess of Orléans, found herself vulnerable to accusations of using magic against Charles VI partly because she was suspected of having previous knowledge of the subject.[34] This fits the pattern of criticising those seen to have influence over rulers, rather than the rulers themselves, and was made easier by Valentina's gender and her connection to a court perceived as containing numerous magicians.

In a similar style, Pierre d'Ailly's *Tractatus contra astronomos* (*Treatise Against the Astronomers*) condemned illicit judicial astrology – but mainly on the grounds of its practical impossibility and consequent dishonesty. This had the advantage of damning practitioners, while portraying powerful patrons as victims of deception. D'Ailly accepted that the planets had powers that affected life and events on Earth, and thus that it could be useful to study astrological charts of major, large-scale events. He also believed that it was potentially instructive to draw up a chart of planetary positions at the time of Christ's birth, to understand the complexion of the heavens at such a crucial moment.[35] Such restraint was echoed by Nicole Oresme, who devotes Chapter 13 of his *Livre de Divinacions* (*On Divination*) to how a good prince should view astrology.[36] The conclusion is that decisions should be made on the basis of political analysis and the advice of loyal, disinterested counsellors. A prince may learn about astronomy and astrology via the instruction of tutors, but should not devote time to technical studies. He should not read Ptolemy himself, nor work with technical tools and tables 'even if the abstention is painful to him and even if the study of such subjects did not take him away from public business'. This is especially the case in time of war and tribulation. The good prince is a patron of true experts, who use careful observations, natural philosophical knowledge and reason to produce answers. Such men are 'sometimes necessary and always to be esteemed and rewarded' – but the ruler should not attempt to be such an expert himself.

Christine's combination of enthusiasm and caution was very necessary, despite Charles V's interest in astrology and alchemy. One factor is that there were powerful and influential voices arguing for the dangers inherent in some forms of astrology, as well as in divination and outright necromancy. The scholar and churchman Nicole Oresme had had the patronage of the king since the 1360s, when Christine was still a small child, and was an informed critic. One of his early treatises was an attack on judicial astrology, written in c.1349. Those arguments were repeated and developed in his longer work *On Divination*, written in c.1366 and attacking all forms of magic.[37] Of course, nobody, including Oresme, thought that judicial astrology was all of astrology. What was condemned was the popular but problematic practice of making personal predictions for clients, concerning coming actions and decisions. As theologians pointed out, human free will meant that such things could not truly be enforced or predicted by the stars. Such forms of astrology were thus both heretical and in danger of being exploited by demons eager to bring about the downfall of unwary humans. Oresme, following Augustine and building on mainstream theologians, distinguished between judicial astrology and acceptable astrology. His own skill in what would now be called astronomy is clear in several of his works, starting with a short commentary on the leading textbook, *On the Sphere*, by Sacrobosco. For Charles V he translated that textbook, as well as Aristotle's *On the Heavens*.[38] The latter was finished in 1377, by which time Christine was resident in Paris with her father, and was apparently favoured by the king and allowed to study Latin and volumes of philosophy – despite the anxieties this caused her mother.

Sadly for Christine, after the early death of Charles V in 1380 the status of Tommaso and his family at court declined significantly. The emergence of rival factions during the period of regency which followed made the quest for patronage complex. Moreover, as the disturbing illness of Charles VI became more obvious, theological arguments, magical practices and political machinations swirled into a dangerous mixture. Charles declared himself sole ruler in 1388 and suffered his first serious episode of mental disturbance in 1392. Chroniclers reported that medical and spiritual causes were suggested for this, but that hostile magic was suspected by the majority of the population.[39] This offered an opportunity for leaders of the rival political factions to hurl accusations at one another – not necessarily of practising such magic themselves, but certainly of employing specialists who could. What is perhaps more surprising is that these accusations had a basis in fact.

As early as 1393, the ambitious duke of Burgundy, Philip the Bold, opened such hostilities against Charles VI's brother, Louis of Orléans. In that year, a magician named Arnaud Guillaume, employed by a group of counsellors, attempted to heal Charles. He used magical rituals contained in a book of supposedly angelic origin called *Smagorad*, but was unsuccessful. When the attempt failed, rumours were circulated that Louis's wife, Valentina Visconti, had used counter-magic, or even caused the illness itself.[40] The rumours were given a sort of credibility by the strong affection which the king showed for Valentina while suffering outbreaks of illness. Finally, in 1396, the duchess of Orléans had to leave the royal court under a cloud of suspicion, against which her husband could not protect her. The fate of Jean (or Jehan) de Bar, one of the king's physicians, suggests retaliation by Orléans. De Bar was accused of sorcery and diabolism in 1398, and the accusations included using invocations against the king and making a demonic image for the duke of Burgundy. A sensational and detailed confession was extracted from de Bar, before he was burnt on the orders of Louis of Orléans.[41]

Belief that magic could both cause and cure illness persisted, and again in 1398 the constable of France employed two practitioners to try to cure Charles. They claimed to be Augustinian hermits, despite their reputation as magicians, and were housed in a royal palace, supposedly with frequent meetings with Louis of Orléans. When their efforts failed, they accused others, including some close to Louis, of using necromancy to thwart them. The ultimate fate of the two 'Augustinians' is unclear; but when two magicians, this time linked to the duke of Burgundy, tried and failed in 1402–03 to cure Charles, they were burnt alive for sorcery. These crises, and others like them, were accompanied by a propaganda war, in which scholarly representatives of both sides produced lengthy treatises detailing the wickedness of magic and the ways in which it was allegedly being used. A high (or low) point was reached in 1407 with the assassination of Louis of Orléans on the orders of Philip of Burgundy's son and successor, Jean 'the Fearless'. The assassination was carried out by physical rather than magical means, but the new duke of Burgundy employed a theologian, Jean Petit, to produce a detailed justification for the act (issued in 1408). Among the claims made were accusations that the murdered duke of Orléans had made use of magic in attempts to harm and even kill the king.

The context for all these machinations was the rise of a disastrous civil war within France. This encompassed the assassination of Jean the Fearless in 1419, on the order of the dauphin, as well as a period of rule by a Burgundian and English alliance. Clearly, if the duchess of Orléans could be suspected of using magic and driven from court, a relatively powerless widow like Christine de Pizan needed to be very careful about displaying any knowledge of divination or heretical forms of astrology. Her success in finding patrons on both sides, while maintaining her status as a woman close to the circles of government, is impressive. It was presumably a factor – together with her obvious skill as a writer – in her selection by Philip of Burgundy to write the biography of the much-lamented Charles V. It was perhaps with some trepidation that Christine responded to Duke Philip's invitation to his court in 1404. That Christine hoped for patronage is shown by the fact that she gave the duke a presentation copy of her recent work, the *Book of the Changes of Fortune*, as a New Year gift that same year. Even so, the request to depict the king must have been intimidating. Fortunately, Christine was able to record that the duke himself gave her guidance on what was required, together with models of historical works to follow. She took care, as she always did, to remind the reader of her humble status; but her strong support for Charles' belief in astrology and astrologers was especially brave at this time.

Christine's earliest works were in the form of poetry and written from a skilled, yet apparently personal, standpoint. When she moved on to her highly original, and very erudite, *Letter of Othea*, she could present herself partly in the accepted role of a loving mother offering advice to her family – even though copies of the work were given to Charles VI, Queen Isabeau and the dukes of Orléans and Burgundy.

Even so, while praise of a king acclaimed as 'the Wise' was relatively safe, display of personal knowledge of astrology, especially by a woman, was considerably more dangerous. Christine's guidance concerning the natures and powers of the planets, given in the very successful *Letter of Othea*, was short and simple. The king, the queen and the dukes of Orléans and Burgundy all received presentation volumes containing copies of this work. Despite its title, it was constructed around a series of a hundred short texts, many in poetic format, with much longer historical, moral and spiritual comments drawing on an impressive range of sources, anchored by an appropriate biblical quotation. It thus offered a compendium of useful knowledge and guidance for life, as well as sources providing instant erudition for readers. No fewer than sixty-one of its hundred 'texts' relate to classical deities and myths, with special interest in planetary gods and goddesses. First to appear is Jupiter (text 6), who provides a model for noblemen. The commentary explains that the pagans gave the seven planets the names of their chief deities, and named the days of the week after them. Jupiter was one of the most

Figure 8.1. Christine as mother and scholar. Paris, Bibliothèque
nationale de France, MS Fr. 12779, f149v.

powerful, and is correspondingly placed in one of the highest parts
of the zodiac. The legendary alchemist, Geber, is cited as a source
for the information that alchemists link Jupiter to the metal tin. In
health terms, Jupiter is linked to the sanguine complexion, corresponding to its positive nature as an encourager of mercy, pity and
compassion, as well as of positive forms of enjoyment. Rather
surprisingly, Saturn (text 8), normally depicted negatively, is almost
equally positive here. His seriousness and slowness provide a model
to be followed, especially by those who have to give judgements. He

Figure 8.2. In this manuscript, owned by the duc de Berry, Christine is depicted as a richly but conservatively dressed woman, using the tools of a scribe. Paris, Bibliothèque nationale de France, MS Fr. 835, f1r.

is linked to Saturday and to the metal lead, and encourages wisdom, as well as sadness.

Venus appears between the two powerful gods and is presented almost entirely negatively, although the information is given that she is linked to Friday and to copper. She is untrustworthy and fickle, and encourages the destructive sins of lechery and lust. There is also a suggestion that Christine subscribed to the view that pagan deities were originally prominent human beings, around whom legends later grew up. This appears when Venus is linked to an ancient queen of Cyprus, who encouraged her followers in multiple love affairs. The Moon is also presented in a negative style. She causes melancholy and 'frenzy' (a medical term with wider meanings than in modern English), which can lead to irrational behaviour. Her constant changes also make her the enemy of constancy and steadfastness. However, this negative depiction is partially balanced by the very positive account of Diana, who was also linked to the Moon. The Moon also has individuals who are subject to her. An innovative feature of the illustrations for this text in the presentation manuscript known as the 'Book of the Queen' (London, British Library, Harley MS 4431) is that the planetary deities are depicted together with their followers or subjects. This is an early appearance of an important element of late medieval and early modern astrology, known as the 'Children of the Planets', suggesting that Christine was very well informed concerning such developments. The main idea was that an individual could identify the planet that played the main part in their nature and life, and could learn about themselves from that. It can perhaps be compared to modern Sun-sign astrology, except that instead of identifying as 'a Virgo' someone might be a 'child of Jupiter'.

This is far from all the wisdom offered in this work, but clearly Christine provided an informed yet individual version of popular astrology and astronomy. The result combines the guidance found in works such as the *Secret of Secrets* and the middle English *Wise Book of Astronomy and Philosophy* with far more scholarly materials.[42] It will already be clear that Christine is far from saying that the fates and actions of individuals are determined by the planets, even while she emphasises that they influence humans. One question that she presumably faced was that of how her father failed to save the life of Charles V or to escape his own fall from grace. An answer is suggested in her *Book of the Mutability of Fortune*, in which Fortune appears like a classical deity.[43] Christine's Fortune is fickle, destructive and disruptive. She cannot interfere with the spheres of the planets and stars, but can disguise, transform, destroy and even cause things below. She

Figure 8.3. The Moon and her subjects London, British Library, Harley MS 4431, f101r.

has power over everything located in finite time. Christine's father, through his devotion to study, achieved mastery of astrology and medicine. He had knowledge of the stars and the celestial regions, and could give advice based on his ability to interpret signs of coming events, such as wars, natural disasters and famines. Christine laments that, as a female, she could not be taught more than 'scraps and flakes, small coins and bits of change' from her father's knowledge – but even these are precious. However, the book makes it very clear that neither the learning of her father nor the virtues of her husband could withstand the capricious blows of Fortune.

Further reflection on the value of astronomy and astrology is given in the *Path of Long Study*.[44] This, too, is written as a personal account of the experiences of its central character, who is a highly educated woman. The experiences in question centre on a visionary dream, and so range far beyond everyday life. The narrator figure travels great distances both on Earth and into the heavens, as well as into the realms of classical legend – which overlap with those above the Earth. Once again, Christine addresses her work to the most powerful individuals in the kingdom: the king; the 'most high princes'; and 'exalted and magnificent dukes'. All are requested to engage with the issues raised by the vision narrated, which is presented as wisdom revealed to the dreamer by one of the classical Sibyls. This strategy enables Christine to support study as a valid occupation for all those suited to it, irrespective of gender, as well as revealing at least some of her own erudition.

For Christine, as for other medieval authors, the Sibyl was in effect a Christian prophetess, who had foreseen the coming of Christ and had prophesied for the pagans. As a figure from both the classical and the Christian worlds, the Sibyl is well suited to lead the dream-Christine to the Earthly Paradise, located on Mount Parnassus and home to the Fountain of Wisdom. This fountain is presided over by Pegasus, the winged horse of classical mythology, and was the abode of great classical philosophers and poets, as well as of the Muses – described as nine powerful women who hold a school there. The first hint that astrology and astronomy are impor-tant parts of wisdom and philosophy comes in the statement that Christine's own father was able to visit this fountain and study in its school.[45] Christine's destiny is to follow the eponymous Path of Long Study, which culminates in climbing the ladder that leads up to the firmament. Knowledge of the Earth and all its marvels is a prerequisite for this, again placing study of the celestial realms at the pinnacle of human achievement. However, as in Christine's previous depiction of astronomical/astrological knowledge, her version is both steeped in classical legend and significantly different from the astronomy taught in standard textbooks. Given Christine's ability to read Latin and the fact that Sacrobosco's textbook on astronomy had been translated into French, this was presumably deliberate.

The first departure from standard terminology is that the Spheres of the Elements, located between the Earth and the Moon according to astronomical textbooks, are blurred with the heavens. The first of these is that of Air and the second that of Ether, which borders the third heaven – of Fire. The fourth heaven is that of Olympus, while the fifth heaven – and the limit of Christine's vision – is the firmament. This, in Christine's version, comprises the seven 'houses' and 'places' (*maisons* and *lieux*) of the planets. The term 'stars' is mainly reserved for the bodies further away, which move when carried along by the firmament (i.e. the 'fixed stars').[46] The Sibyl expounds the names and powers of each planet, together with the basics of their (apparent) movements around the Earth, and the locations of the main groups of stars. A surprise follows, which is that Christine uses some selected astrological terminology. The 'properties' of each planet are explained, while their actual effects are linked to their relationships to one another – specifically to Oppositions between planets and to their changing positions. Solar and lunar eclipses are also identified as important, as are the movements of the Sun both around the Earth and within the zodiac. At this point the narrative is halted and the dreamer says that no more will be said about the Sibyl's instructions on such topics, since 'it is not appropriate for me to meddle in judgements of such knowledge as I did not learn astrology in any school'.[47]

The tone then shifts to direct description of what was seen in the vision, and the dreamer marvels at the order and stability of the spheres that make up the universe as ordained by the Creator. The technical terminology deployed in this more astronomical passage is fuller and more confident. The axis and poles of the universe, and the rotation of the spheres around them, are set out without apology or hesitation; even the technicalities of 'great circles' and of the key points in the Sun's circuit of the ecliptic are briefly covered (though those terms are not used). The same confidence is shown in statements like: 'I saw the twelve signs in their places, in the furthest part of the sky, set in four groups and in two sets of six.' This description assumes some astronomical knowledge, and prepares for the statement that the signs are also houses for the planets. Indeed, the Sibyl reveals 'the whole mystery of the planets', including the zodiac signs in which each is especially powerful (their Exaltations).

However, that is the limit of what the dreamer feels it appropriate to reveal, before moving on to a description of the Sun's splendour and its 'chariot', and of how the Sun and Moon, in particular, create cycles of time through their movements.[48] The revelation of the universe is so intense that the dreamer is moved to an almost ecstatic state of praise for its Creator, before being brought back to a sense of her physical limitations by the Sibyl. This part of the work ends with the Sibyl's statement that the universe perceptible to Christine has as its outermost boundary the Crystalline Sphere, beyond which is the abode of saints, angels and, ultimately, of God Himself. In other words, this dream vision gives more technical details of astrology than the biography of Charles V, while repeating that study of the cosmos and its workings is eminently justified by the intimations of the divine to which it leads.

This emphasis on the political and spiritual value of astrology is echoed in Christine's *Book of the Body Politic*, which goes so far as to say that it is appropriate for skilled artisans to be taught measurement and proportion through study of geometry; but astrological knowledge is reserved for philosophers and the princes who consult them.[49] This treatise suggests a rather radical interpretation of the idea that people can and should be classified by their dominant abilities and characteristics. Christine puts forward the view that, in an ideally ordered society, roles should not be allocated simply according to social rank: such abilities and characteristics should be taken fully into account. As noted above, she espouses the established view that deep study of astrology goes beyond what is appropriate for a prince, and that this should be left to philosophers. Nevertheless, she gives practical examples of the political and military benefits to be gained from astrological and astronomical knowledge. One such is the story of the 'philosopher' Thales, which Christine attributes to Aristotle. Thales was criticised by his family for devoting himself to study, rather than advancing his career, and so used his skill in astrology to predict future changes in the olive harvest. His consequent investment enabled him to make an impressive profit and so silence his relatives. There is also a suggestion that Julius Caesar could have escaped the assassination plot that caused his death if he had only listened to the astrologer whom Christine calls 'Spiromia'.[50] Christine goes on to provide early examples of the trope concerning the power of accurate understanding of eclipses. Both 'Supplicius Gallus' and Alexander the Great are said to have maintained the morale of their own soldiers and achieved important victories through such knowledge.[51]

This evidence suggests both that Christine placed an extremely high value on astronomical and astrological study and that she was very skilful when it came to displaying her own knowledge in acceptable ways. The reputation of her father is upheld, although the existence of fraudulent astrologers is acknowledged. Knowledge of the heavens is of the highest value both for the practical guidance it offers to those who can make informed use of it and for the insights it can give into the splendours of heaven itself. At the same time, Christine emerges as an early adopter of a fashionable and increasingly popular form of astrology, centred simply on the seven planets and their interactions with the signs of the zodiac. The rise and spread of this form of astrology does much to explain the fifteenth-century interest in the treatise of Fendulus, which provided an apparently authoritative and attractive introduction to this subject matter. The information given by Christine was securely set within texts which guided the reader's response, and which in many ways concealed the depth and novelty of her material.

Some examples will show the rapid spread of this newly popular form of astrology, both geographically and to new groups of consumers. Of a quality comparable to the luxury books circulating in the French royal court is the impressively illuminated manuscript made for Francesco Sforza and now in Modena's Biblioteca Estense, as MS Lat. 209. This is rather misleadingly known as *De sphaera* (*On the Sphere*), although it is not an astronomical textbook like those of Sacrobosco or Grosseteste, and has no title and very little text in its current form. After an introductory group of diagrams displaying the structure of the universe and the mechanics of eclipses, it moves into a set of beautiful images of the seven planets and their children, accompanied by short, poetic texts. Here, once again, Saturn is a severe, but ultimately just judge; meanwhile, the Moon is less negative than in Christine's account.

Vernacular versions of this desirable knowledge spread not only in French, but also in Middle English, German and Middle Dutch – a striking example of which is the manuscript now Wolfenbüttel, Herzog August Bibliothek, Cod. Guelf. 18.2 Aug. 4°.[52] This is a treatise on the nature of the universe, which puts emphasis on the powers of the planets and the signs of the zodiac within a brief encyclopaedic coverage of natural philosophy.[53] It is most remark-able for its illuminations of the planets, which are comparable to those in Christine's presentation manuscripts and in the Sforza *Sphere*. Its depiction of the Moon is even more negative than that in Christine's work, and links the Moon to the idea of fate or Fortune.

Figure 8.4. The Moon as Fortune. Wolfenbüttel, Herzog August
Bibliothek, Cod. Guelf. 18.2 Aug. 4°, f123r.

The new technology of printing made it possible for illustrated
woodblocks to be used to spread more affordable versions of the information in German-speaking cities, such as Salzburg and Basel. Most
of these texts and images are later than Christine's work, which once

again emphasises her powerful combination of innovation and caring instruction.[54] A final point of importance is that Christine gave her support to astrology alone, out of all the forms of prognostication and prediction represented in the library of Charles V. There is nothing in her works to suggest engagement with palmistry (chiromancy) or even the fashionable art of geomancy. Alchemy, despite its popularity with fifteenth-century rulers and scholars, is also absent. The effect is to present astrology and astronomy together as one wide, but unified field of knowledge, with prestigious origins in the ancient world. The value of such knowledge is linked not just to its origins, however, but also to more modern developments and practical applications. This is likely to be evidence of Christine's own views and this was a bold stance, especially for a woman.

However, Christine clearly judged the interests of her patrons and potential critics very astutely. In her hands, the science of the stars was worthy of respect for the supernatural knowledge that it offered – but that knowledge was ultimately a gift from God. Readers would not know from Christine that claims and counter-claims of necromancy, magical murder and heretical demon-summoning were part of political discourse. In her works, knowledge of the stars can provide useful guidance without trespassing on the concepts of free will and individual responsibility. Astrological knowledge allows skilled individuals to gain knowledge of God's universe and understand the celestial influences acting upon them and others. Such knowledge can be a powerfully helpful for those responsible for making decisions that will affect the lives of many, as well as for those simply wishing to live a healthy life. It is not surprising that overt divination is not recommended by Christine – although the next chapter will show that at least one form of such divination, that of geomancy, was flourishing in her lifetime and later. Geomancy was another 'art' which inhabited a complex border zone between natural science and divination, raising questions about how closely magic was defined and policed in the corridors of power – and by whom.

9

—◇—

DIVINATION AND POLITICS

The Rise of Geomancy

Geomancy makes an appropriate concluding subject for this book, due to its popularity in both the late middle ages and the Renaissance, including among those exercising high levels of power. The level of interest is attested to both by the sheer number of surviving manuscripts in which it appears and by its presence in library catalogues from the period. Not least among these is the library list of King Charles V of France (discussed in Chapter 8), which included no fewer than nineteen entries for geomancy. Moreover, a recent study of central European manuscripts containing magical texts has concluded that works on geomancy, of varying lengths, are to be found in almost all of the manuscripts examined. They have tended to be missed or wrongly identified by cataloguers unfamiliar with the subject, thus perpetuating their lack of visibility.[1] The shorter texts are especially vulnerable to this problem, since they are often placed between two longer and better-known works.

Another striking characteristic of geomancy has been its near-invisibility, until recently, to historians. Even Lynn Thorndike paid scant attention to geomancy in his impressive *History of Magic and Experimental Science*. Thorndike called geomancy a 'superstitious art' and gave only a brief statement that it worked by marking down random points and then interpreting them via both sets of tables and 'astrological constellations'. He paid more attention to it as a problem for twelfth-century theologians and commentators on magic and divination than as an actual practice. Nevertheless, he provided a list

of the main translators whose names are linked with geomantic treatises in Latin, and commented that texts occur frequently in surviving
medieval manuscripts. He gave a list of twenty-three such manuscripts identified in the Bayerische Staatsbibliothek in Munich as an
example of this phenomenon.[2] The most important study of medieval geomancy remains that of Thérèse Charmasson, published in
1980, which listed almost 150 manuscripts.[3] However, new archival
research, especially by Alessandro Palazzo, has added further to
Charmasson's total.[4]

Historians of divinatory procedures in the Islamicate world have
added considerably to knowledge of the early history of geomancy,
which has supported further work on Latin geomancy. It is now clear
that geomancy, despite the claims of some medieval, Latin translators, originated in the Arab Empire.[5] No early manuscripts in Arabic
are known to survive, but references to geomancy, named *khatt al-raml*
or 'science of the sand', are known from at least the tenth century
onwards and assume that the practice was already well established.
The silence of Latin texts concerning this true origin was broken only
by the anonymous, twelfth-century text known by its opening words
as *Estimaverunt Indi*. This was also the only geomantic work to be
singled out for condemnation in 1277, when Stephen Tempier,
bishop of Paris and chancellor of the expanding University of Paris,
issued his list of forbidden works.[6] Tempier placed at least this version
of geomancy, believed in medieval Europe to be the oldest text, in the
company of works on necromancy and demonic magic. The condemnation may also be responsible for the rather curious manuscript
history of this text, since it is widely found in surviving collections of
geomantic works, but often only as excerpts or without clear identification.[7] This fragmentary and anonymous presence suggests that its
knowledge was desired – but in a deniable form.

A distinctive feature of Latin geomancy is that astrological
concepts and procedures are already present in the earliest texts,
although they are much rarer in comparable Arabic treatises. This

suggests that the translators also engaged in editorial activity, presumably to make geomancy fit better with the wide range of scientific, astronomical, medical, magical and divinatory works that they were making available in Latin versions. As the texts and manuscripts discussed in the preceding chapters have shown, the science of the stars held a high status in Latin Europe. The accepted powers of the celestial bodies provided valuable explanations for otherwise unacceptable claims. Latin geomancy is thus found in manuscripts as a hybrid practice, although it can also be seen as a superior form of divination. Part of the attraction may have been that the geomancer's body and spirit formed a personal connection between the Earth below and the celestial spheres above. This made the reception of guidance concerning a problem or question both a (potentially) private experience and an elevated one. For instance, Gerard of Cremona, who produced one of the earliest Latin translations, stated that geomancy drew on 'astronomical and astrological truth', while producing a response to a question more clearly than could often be done by purely astrological methods. It was also helpful, as historians have pointed out, that geomancy offered a less technical access to such knowledge than horoscopic astrology, since no planetary tables or mathematical calculations were involved.

That claim of relative speed and simplicity might suggest that geomancy was favoured by those who could not afford a full consultation with a professional astrologer. Indeed, the appearance of versions in most medieval vernacular languages demonstrates that it was read and used by those who were literate, but not expert in Latin. However, its early audience in Christian Europe appears to have been composed of scholars who were also interested in astrology. Perhaps surprisingly, geomancy was already being practised and taught at the cathedral school in Hereford in the late twelfth century. Evidence for this is provided by Simon de Fresne, a canon of Hereford – and a poet – who boasted to Gerald of Wales in a poem about the impressive range of expertise available at the cathedral. The poem can be dated 1195–97, when a group of scholars was brought together by Bishop William de Vere. Of all the subjects expounded in the poem, it is astronomy that is presented most enthusiastically – and it is already associated with geomancy. Simon claims that 'Here the geomancer forms fifteen figures' and is thus able to tell any man or woman whether his or her wish will be fulfilled. Simon himself shows considerable geomantic knowledge, since he gives the names of three geomantic figures (People, Prison and Head of the dragon, or Populus, Carcer and Caput draconis). Simon also knows that a geomantic reading involves looking at the interaction between any given figure and its location in the full geomantic 'tableau'. There are fifteen or sixteen such positions, depending on the version followed, and the first twelve were treated in Latin geomancy as equivalent to the twelve astrological houses (distinct from the signs of the zodiac in geomancy, just as in astrological texts).[8] Simon's reference to the formation of fifteen figures shows that he was familiar with this procedure.

Further testimony to its popularity among scholars, although of a more critical sort, is given by Thomas Aquinas in his letter-treatise *De sortibus* (*On Casting Lots*). Aquinas acknowledges that lot-casting can be both informative and acceptable, when it is limited to enquiries concerning natural phenomena over which the planets can have significant influence, such as the weather. However, individuals who seek to know things that go beyond this category are in danger of practising necromancy and falling into the clutches of demons. A sort of border category includes those who 'seek knowledge of things hidden' by observing physical signs. Geomancy is described in some detail as an example of types of divination by lot-casting, which can be either licit or demonic. It becomes dangerous when practitioners claim that 'all human actions result necessarily from the stars'. The threshold is therefore crossed when practitioners claim that the hand movements which produce the marks that are to be interpreted are themselves guided by the stars. As Aquinas puts it:

if a geomancer marks out points in the dust they claim that his hand is moved according to the power of the heavens to such an extent that the number of points which appears is in accord with the disposition of the heavens ...

The claim is that such celestially directed casting of points will reveal both the disposition of the planets and the likely outcome of a question asked under such influences. Needless to say, Aquinas' conclusion is that the argument is false.[9] Even so, the knowledge Aquinas reveals here suggests that he has taken time to learn about the subject and believes that it needs informed consideration. The practice was also discussed in detail by Michael Scot both in his own *Liber geomantiae* of c.1230 and in Part 3 of his wide-ranging *Liber introductorius*. Michael Scot, as befitted an expert adviser to an emperor, was not only concerned to discuss the theoretical issues raised by geomancy, but also showed practical knowledge.[10] It thus seems likely that geomantic consultations were practised at the imperial court.

Moreover, there is considerable evidence that geomancy enjoyed ongoing popularity among the rich and powerful in the fourteenth and fifteenth centuries. The king of France was not the only member of a royal dynasty to possess luxury copies of geomantic texts. A manuscript now in Oxford's Bodleian Library appears to have been produced as a special gift for Richard II of England. It has been described as 'Richard II's own book of divination', but is dominated by a beautifully produced work on geomancy.[11] The preface to this text claims that it was composed for the king, who had expressed a 'special request', and that it summarised the rules and precepts of established authorities in this art, 'up to the year of our lord 1391, in the month of March'.[12] The text echoes Gerard's claim of the relative simplicity of geomancy – and is specially designed to make it possible for a user to ask and answer a question without needing to consult a professional. The manuscript contains an impressive illumination of a king, presumably Richard himself, being presented with the book, and the painting appears to be the work of a professional artist based in London. The geomantic text is richly illuminated, with each geomantic figure represented in an individual painting, placed beside a tabular presentation of its name, characteristics and significations (Figure 9.1).

Still more impressive, and in the classicising style of the Italian Renaissance, is a manuscript now in the National Art Library, part of the Victoria and Albert Museum, London. This manuscript appears to have been created in northern Italy in the fifteenth century, and is written in a beautiful version of 'humanist' script, intended as an imitation of Roman script. It opens with a flattering image of the reader as a classical philosopher/scholar, who is thinking deeply while consulting a text of geomancy in which versions of the figures are suggested, as well as manoeuvring an armillary sphere (Figure 9.2). The status of both book and reader is further stressed by the fact that the outer border of this folio is

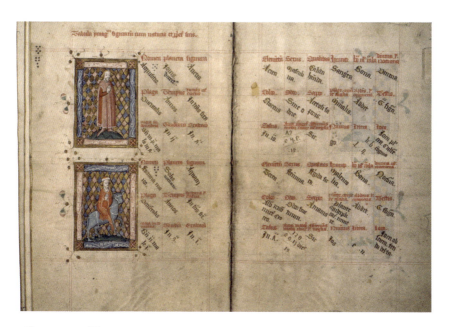

Figure 9.1. The geomantic figures Acquisitio and Fortuna major. Oxford, Bodleian Library, MS Bodley 581, ff15v–16r.

stained with 'imperial purple' and contains depictions of the planets as classical deities. The chariot of the Sun and a crowned Moon deity can be seen in the bottom corners, flanking a shield containing arms, presumably of the original owner of the volume. The scribe was Bartolomeo Sanvito (c.1433–c.1511) – one of the greatest practitioners of the humanist script fashionable among Renaissance patrons and scholars, who worked in both Padua and Rome for a series of prestigious patrons.[13]

As Hilary Carey has shown, it seems that geomancy was also practised in the court of Humphrey, duke of Gloucester, and his second wife, Eleanor Cobham. Given Duke Humphrey's patronage of fashionable authors and valuable books, it is unsurprising that a practice as favoured as geomancy should be represented in his collection. What is less usual is that Duchess Eleanor herself seems to have been

Figure 9.2. A geomancer in his study. London, National Art Library, MSL/1950/2464.

directly interested in this art. The evidence is preserved among surviving fragments of texts written by Roger Bolingbroke, principal of St Andrew's Hall, Oxford (1438) and a scholar and astrologer in the service of Duchess Eleanor. That closeness to the duchess was to cost Bolingbroke very dear, as he was condemned and executed for treasonous necromancy, supposedly carried out for Eleanor. This was part of the scandalous events of 1441, when Eleanor, Bolingbroke, the learned physician Thomas Southwell and Margery Jourdemayne (the 'Witch of Eye') were found guilty of necromancy, treason and heresy in their supposed attempt to bring about the death of Henry VI through complex ritual magic. Geomancy was not part of these accusations, but Eleanor was known to be interested in astrological medicine and other unfeminine subjects, and was the recipient of Roger Bolingbroke's treatise on geomancy. The full text of this work does not survive, but it was apparently highly regarded, since John Argentine, provost of King's College Cambridge and royal physician, preserved extracts from it. It was written in English, making it possible for the duchess to read it herself, and was dedicated to an 'esteemed and most revered lady' who seems to have wished to understand the basics of geomancy. The name of this powerful and admirable lady is not given, but Duchess Eleanor is the most likely candidate.[14] It is worth noting that, if Eleanor and her advisers wished to discover the likely date of the king's death, which was part of the accusations against them, then geomancy (especially in the expert hands of Bolingbroke) could have provided an answer. Margery Jourdemayne, previously convicted of witchcraft in the form of potion-selling, would not be needed for such intellectual activities.[15] Yet Duchess Eleanor only confessed to consulting Margery – and then in the hope of becoming pregnant. The elaborate making and baptism of a wax image of the king, with which the group was charged, was a well-known form of ritual magic, but clearly exceeded the scope of geomancy.

At this stage it may be useful to give an outline of how geomancy was carried out in practice. It would be satisfying to use the work dedicated to Duchess Eleanor; but as that is not possible, the following account will be based on a Middle English treatise, which is available in a modern edition.[16] This Middle English version exists only in a fifteenth-century manuscript, which is now Oxford, Bodleian Library, Ashmole MS 360, but is representative in the form of geomancy it presents. Expert knowledge is shown by the facts that some of the materials which it includes remain in Latin and that it is linked to the earlier work of Martin of Spain. Martin of Spain's Latin text is dated to c.1319 and has the advantage of impeccable credentials. Martin of Spain himself is described in the manuscript as an abbot of Cernat, in Burgos (Castile), and a skilled physician; moreover, he wrote for another abbot, named Archimbold of St-Astier in Périgueux. Analysis of Martin's text suggests that he drew on a range of older treatises. It is not found in very many surviving medieval manuscripts, but has been identified in two manuscripts, one of c.1400 and one of c.1600.[17]

Like many other geomantic treatises, this version is a practical handbook. It guides the practitioner, whether expert or beginner, through the process; and, like many others, it offers ready-worked examples of how to provide an interpretation and 'judgement' in response to given questions. In giving ready-made questions and answers, such texts have something in common with the divinatory 'books of fortune' used by Matthew Paris and discussed in Chapter 3, although a full geomantic consultation would take longer and require considerably more knowledge of astronomy and astrology. Despite its emphasis on practicality, the manuscript opens in scholarly style, with a few lines in Latin establishing the origin and meaning of the name 'geomantia'. The switch to English accompanies a switch to direct instruction, with guidance on preparing the sand, if it is to be used as a medium for casting the figures, and on the tools for making the dots which comprise the figures. This distinction leads to a slightly odd mixing of languages at the end of the introduction. The instruction that it is best to use one's finger for casting the dots is given in

English, while the theoretical explanation that this allows the 'guiding intelligence' of the celestial spheres to guide the hand of the geomancer is in Latin.

With the fundamentals thus established, the text proceeds to the geomantic casting itself. Geomancers should first focus intently on the question being asked, and hold it clearly in their minds, while also asking for God's help. Then, still holding the question clearly in mind, they should make four sets of four lines of dots or 'points' (sixteen lines in all). It is important not to look at or think about the number of points being cast in any line. Once all sixteen lines have been produced, the basis for the rest of the process has been put in place. The next stage is to produce the first four geomantic figures, known as the 'Mothers', from these lines of points. Reading from right to left (in the Arabic manner) the geomancer strikes out pairs of points, until only one pair, or a single point, remains at the end of the line. For instance:

· ▪▪▪▪▪▪▪▪▪▪▪▪

In this way, the first four lines will produce four single or double points, one below the other. There are sixteen possible combinations, each of which makes up one geomantic figure.

The full set of sixteen possible combinations of single and double points, constituting the sixteen named geomantic figures, is clearly illustrated in the seventeenth-century printed edition of Henry Cornelius Agrippa's work (Figure 9.3). For example, the three figures named in Simon de Fresne's poem above (plus Via – the Way) would look like this:

Populus	Carcer	Caput draconis	Via
· ·	·	· ·	·
· ·	· ·	·	·
· ·	· ·	·	·
· ·	·	·	·

Each set of four lines, treated as explained above, thus produces one Mother figure, and all four of these are drawn beside one another, again from right to left. The set of four Mother figures shown above will be used as an example.

The next stage is to produce the 'Daughter' figures from the Mothers. This is done by reading across the lines of points, again working from right to left. In the example above, this would produce first a single point, then a double, then a single and finally another double. Arranged vertically (as is standard for geomantic figures), the result looks like this:

.

. .

.

. .

This is the figure named Amissio (Loss). The same procedure is then followed for the remaining three lines of the Mother figures, to produce all four Daughter figures. These are added into the growing geomantic 'tableau' beside the Mothers, still working from right to left. In the example given here, the other Daughters would be: Fortuna minor; Fortuna minor again; and Cauda draconis (Dragon's tail). With these in place to the left of the Mothers, a line of eight figures side by side forms the upper part of the tableau.

A different procedure is now used to produce four more figures from these eight. Each of this third group is generated from a pair of the figures in the top row, by counting whether the points in their joint rows are even or uneven in number. For instance, in the tableau being developed here, the first Mother figure, Via, has a single point at the top, while the second Mother figure, Caput draconis, has two, adding up to three. This uneven total, treated as in the first stage, means that the top row of the new figure will contain one point. By the same process, the second row will have two points, as will the third and the fourth. This means that the first figure in the second tier will be Letitia (Joy). Similarly, the second will be Carcer, generated from Carcer and Populus. The first pair of Daughters, Amissio and Fortuna minor, will produce Conjunctio, and the final pair of Daughters produces Albus. Each of these is entered into the tableau, placed below the two figures from which it was generated, resulting in a row of four.

Figures 9 and 10 are used next, following the process just described above, to generate the first of the two figures that will make up the third row of the tableau. This will have three pairs of points above a single, making up the figure called Tristitia (Sorrow). In the same way, figures 11 and 12 produce a figure whose four rows contain respectively two, one, two and two points – this is Rubeus. At this point the end of the casting of the figures is nearly in sight. Figures 13 and 14 are next used in the same way to produce figure 15, which will be Acquisitio (Gain), having rows of two, one, two and one points. Finally, figure 15 is paired in the same way with the first of the Mothers to produce the sixteenth figure. For us, this will sadly be Amissio (Loss), having rows of one, two, one and two points.

With all sixteen figures now cast and placed in a triangular tableau, the text proceeds to expound the names and significance of all of them (it is not necessarily the case that all sixteen will come up in any one casting). The simplest way to illustrate the figures is to give a diagram. Figure 9.3 reproduces an illustration printed in a copy of Robert Turner's English translation of Henry Cornelius Agrippa's authoritative account of geomancy. It gives the names and shapes of all the figures, together with the planet to which each is linked in geomantic interpretation – information that is central to the astrological procedures included in Latin geomancy.

The next step is to expound the astrological meanings and affiliations of the figures, starting with the explanation that each point in each figure represents a star or star group, while each complete figure is linked to a zodiac sign (some signs having two figures). In the case of the seven planets, each has two figures, one more positive

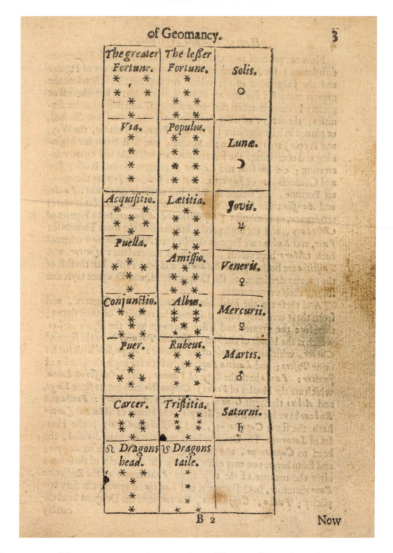

Figure 9.3. The geomantic figures, from Henry Cornelius Agrippa, *Fourth book of occult philosophy, Of Geomancy* trans. Robert Turner, printed by J.C. for J. Harrison, London, 1655, p.3. London, Wellcome Collection.

in meaning and the other more negative (as in Figure 9.3). For Saturn, Jupiter, Mars, Venus and Mercury, the more positive figure indicates the planet moving 'direct' in its normal fashion; the more negative figure indicates 'retrograde' or reverse movement. Caput and Cauda (Head and Tail of the Dragon) are linked respectively here to Conjunctions of Jupiter and Venus (the fortunate planets) and Saturn and Mars (the malefics). The author of the Ashmole treatise is then at pains to explain how the astrological qualities of the planets and signs underlie the relevant figures. This in turn supports the idea that the geomantic reading is equivalent to an astrological reading.

An important question relates to the complex issue of the placing of each figure in the geomantic triangular tableau, and how the positions equate to the 'mundane' houses used by astrologers. There are twelve of these, each covering one part of the nature and life of the person asking the question. They are calculated in relation to the querent's location, starting at the eastern horizon, and so (unlike the zodiac signs) remain fixed in position. This means that the signs and planets rotate through all of them in every twenty-four-hour period, resulting in important interactions between the houses and the celestial bodies placed in them at any given time. Martin of Spain's text does not deal fully with this at this stage, but raises it as an addition to the discussion of the levels of positive and negative meaning attributed to the figures named Tristitia and Letitia (Sorrow and Joy). The argument is made that, although Sorrow is usually negative in meaning, this will be altered if it is placed in a house signifying oncoming change or an ending to something – in these cases the message is that a sorrow will soon be over or will change to something more positive. The reverse is the case for Joy.

This exposition opens the way for discussion of a series of practical questions and answers, related to individual figures, where house position is an important issue. Interestingly, these are very practical questions, mainly dealing with business affairs, rather than personal and emotional matters. For instance, one asks whether the money issued by the king of Castile in 1307 will be stable; and the response is that it will be. This question was clearly checked, since a Latin note points out that this is indeed what happened. Several other questions relate to transactions involving the payment of debts and investments, and whether the promised money will actually arrive. Considerable detail is possible, since methods are given for deriving dates, quantities and other such information from the figures. Weather forecasts are also possible, since the figures, like the planets and the signs, are linked to elements and to times. A large quantity of detailed information is given, requiring considerable effort of memorisation on the part of the geomancer. The fundamental message seems to be that the geomancer should decide at an early stage to which of the houses the specific question relates. This in turn will identify the geomantic figure(s) that are most significant; while the interaction between figure and house position, and all their layers of meaning, will be especially important.

It is only on folio 26v, after several more discussions of the elemental, planetary and zodiac meanings associated with the figures and how they can be understood in relation to different types of question, that detailed descriptions of the houses themselves begin. For instance, the first house signifies the questioner, the questioner's place of residence and immediate family members. It can also signify an individual's physical health or intentions – and whether that person will carry them out. Being the first house, it can also signify the beginnings of undertakings. For these reasons, the geomantic figure placed in this house will tell the geomancer a lot about the questioner. We also learn that the figures linked to Mercury (Conjunctio and Albus) are especially strong in this house. The text then moves on to the second house, which is also expounded in some detail. Its main signification relates to the achievements, gains and possessions of the questioner, though it can also pertain to a colleague or a woman with whom the questioner has a secret or close relationship.

The Middle English translation breaks off at this point, but the text is continued in Latin. The reason for this change is unclear, but the effect is that the next topics, which are technical procedures based on methods for interpreting horoscopes, are presented only in Latin. This includes coverage of the rest of the twelve houses, and how they interact with individual geomantic figures. Helpfully, Henry Cornelius Agrippa later summed up the relationship between placings in the geomantic tableau and astrological houses straight-forwardly: in effect, the first twelve figures in the triangular tableau are to be understood as occupying the twelve houses.[18] The count starts at the top right corner of the tableau and moves from right to left along each successive line.

The final, and even more technical, topic is that of astrological aspects. In a horoscope this involves calculating the angle of separation between two planets to gain further information about the strength, weakness and mutual influences of the planets involved. For instance, if planets are located in signs or houses facing one another across the circle of the zodiac, they are said to be 'in Opposition' and have a negative interaction. In this case they will be separated by six signs. Equally, if they are close to one another, they are 'in Conjunction' and have a mainly positive relationship. This will normally mean that they are in the same sign and house. The same idea was applied in geomancy, and the calculation simply involved counting how many positions of 'houses' separated two figures.

Sadly, the full set of sample questions, with worked responses and interpretations, is also missing from the original Middle English translation. This absence was clearly problematic, since a later scribe has added shorter entries, in English, for all twelve houses and the main types of question they cover. This begins on folio 44v and ends on folio 47v. The result of all this is that a full account of geomancy, based on the work of an eminently respect-able author/editor is produced, even though the text would not be fully usable for a reader unfamiliar with Latin. At this point it should be said that the appearance of the geomantic text is good, but not at all luxurious. It is currently bound in Ashmole 360 with a collection of other short, mainly technical works, which date from the thirteenth century onwards. 'Geomancy' is one of the latest sections, and has no clear evidence as to the identity of its scribe or original owner. The fact that the English translation is incomplete may suggest that a hoped-for patron lost patience or interest; but the work to complete the coverage of geomancy, carried on into the sixteenth century, shows that the text continued to be of interest. The combination of Latin and English, coupled with the level of detailed information and the emphasis on relevance for physicians, suggests that the treatise may have been the property of a practitioner with a clientele among the urban middle class, whose own Latin was not entirely confident.

One example of such a professional (albeit in a somewhat later period) is the London-based astrologer and geomancer Simon Forman. He was the author of a treatise on geomancy, the *De arte geomantica*, which survives as Oxford, Bodleian Library, Ashmole MS 354. Forman himself owned a separate treatise, which also now forms part of Ashmole 360. This is the fifth section of the composite manuscript, which has attracted attention because it contains the text *On the Astrolabe*, attributed to Chaucer. Forman suffered some discrimination and unhappiness because he could not afford to take a full university degree, but nevertheless achieved considerable wealth and popularity for his successful medical and advisory consultations. His casebooks survive and have been digitised. They demonstrate that, while he made most use of astrology, this was frequently combined with geomancy.

One example is shown in Figure 9.4. This page shows geomantic tableaux on the right, containing fifteen figures (the sixteenth was not part of all methods). Forman has dated the consultation to May 1599, and noted that the reading was made for Mrs Waller, and that it concerned the life and death of her husband.[19] A better-known client, who consulted Forman several times, was the poet and author Emilia Lanyer (or Lanier). Forman's interactions with her are

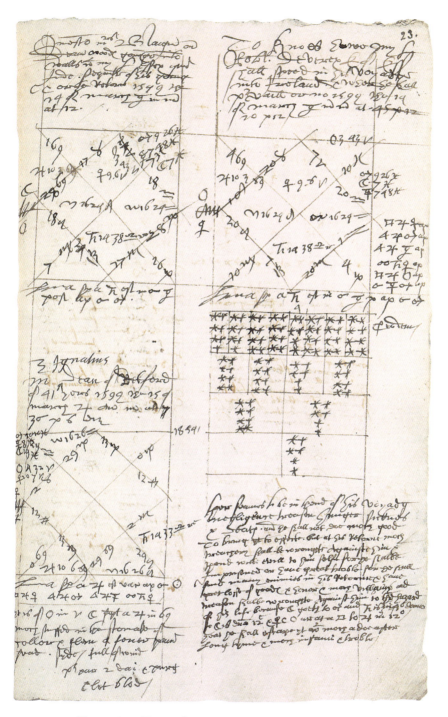

Figure 9.4. Forman's geomantic chart for Mrs Waller.
Oxford, Bodleian Library, Ashmole MS 219, f66r.

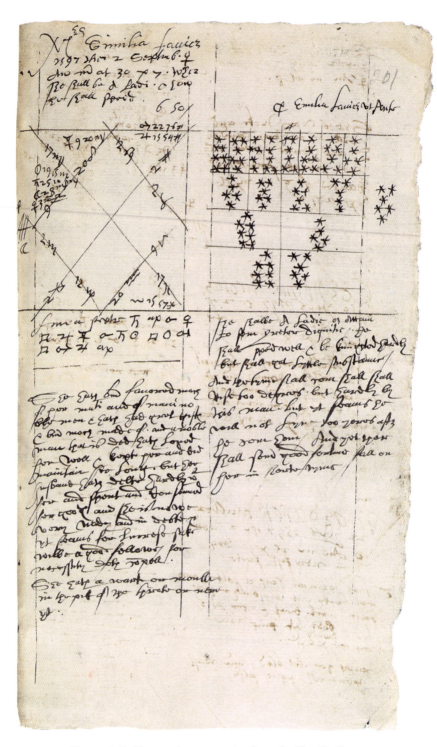

Figure 9.5. Forman's geomantic chart for Emilia Lanyer.
Oxford, Bodleian Library, Ashmole MS 226, f201r.

representative of his attitudes towards female clients, as well as of his geomantic and astrological expertise.[20]

In Figure 9.5, Emilia Lanyer's name appears prominently at the top of the page, with a full geomantic tableau in the right-hand column. The sixteenth figure is included here, and repeated in the margin as an overall conclusion to the reading. It is Acquisitio or Gain, and a very positive sign. This presumably contributed to Forman's view that Lanyer will 'attain greater dignity' and 'have good fortune soon'.[21]

Forman's notes on Emilia Lanyer are very intrusive, containing speculation on her sexual history, comments on her physical appearance and self-serving enquiries as to her sexual availability. This is a reminder of the very personal nature of a consultation with an astrologer or geomancer, and helps to suggest one reason why simpler forms of prognostication and lot-casting – which could be carried out in private, without any need to hand over personal information (or cash) to a professional – were so popular. It also helps to explain the amount of work that was put into making the geomantic volume prepared for Richard II of England, which could be used without needing a professional geomancer. The manuscript – Oxford, Bodleian Library, MS Bodley 581 – has been discussed by several historians, mainly because of the introductory texts that preface the volume itself. These express a desire to support Richard in becoming a good and wise king, as well as offering helpful 'secrets' supposedly derived from the advice given by Aristotle to Alexander the Great.[22] That prologue suggests that the carefully edited and beautifully illuminated geomantic collection, which occupies folios 9 to 89 of the manuscript, is also intended as a source of guidance. The king was making known his desire to be free of what he perceived as overbearing counsellors, but his choice of advisers caused serious concern. For those reasons, an attractive book, which could offer immediate and private insights into both personal relationships and coming events, could have been a very useful gift.

Sadly, however, the manuscript's pristine state suggests strongly that the king made virtually no use of it.

The title *Libellus geomancie* (*Little Book of Geomancy*) is somewhat misleading, given the length and wide range of its contents. As previous analyses have shown, it draws on most of the highly regarded geomantic treatises, and arranges the information in clearly labelled tables. It opens with a detailed exposition and justification of geomancy, arguing that it is not an illicit art and that it can produce 'judgements' which are true. On folio 11r it proceeds to the grandly titled *Book of Judgements*, which begins with an alphabetical table of topics for possible questions, linked to the relevant houses. The next section takes the form of another table, also alphabetical, and bringing the planets into the equation. The most visually attractive part of the manuscript follows on folios 15v to 23r. This introduces each geomantic figure in turn, with just two figures to each opening. For easy interpretation, each figure is accompanied by a personification that sums up its dominant qualities (see Figure 9.6). Those are followed by tabular presentation of the full set of qualities and meanings that the figure embodies. A somewhat surprising characteristic of this catalogue of the figures is that the patterns of points which comprise each one are illustrated only on a small scale. This is presumably because they are depicted on tabs which act as bookmarks, enabling a user to find a desired figure immediately. They are protected by being placed within the margins of the folios and made visible by cut-aways. Larger tabs would make finding the appropriate one less easy.

What is more surprising is that the process of casting the points and producing the figures, explained above, is not included here. It seems that the king was not expected to carry this out himself – which raises the problem of whether the apparent aim of giving the king a source of advice that he could consult directly could ever have worked in practice. However, the rest of the manuscript offers an answer to this question. This is that folios 24r to 75r set out twenty-five key questions, with worked answers attributed to 'Abdallah'. If they did not cover what was needed, then folios 75v to 76r offer a further alphabetical list of topics for questions, followed by further full examples of geomantic enquiries on folios 76v to 87r. Even more helpfully, folios 87v to 89r provide a special *Rosary of King Richard*. This is a list of thirty-five possible questions, with tables illustrating the geomantic figures which provide positive, negative or neutral 'judgements' in relation to each. No instructions are given for the practical application of this information, but the structure of the manuscript itself offers a possible answer.

The image in Figure 9.6 illustrates the last pair of geomantic figures, as presented in the manuscript. It can valuably be compared with the first pair, shown in Figure 9.1 above. The personifications, names, planets, zodiac signs, elemental qualities, direct or retrograde movement and so on can be seen for each. At the left can be seen the cut-in tabs, marked with the appropriate patterns of points (and showing very little sign of use). As these images show, the most effective way of using the book is to start at the back – where the table of contents will also be found. It is from this position that all sixteen tabs, for all sixteen figures, can be seen. By contrast, as in Figure 9.1, the first opening, for Acquisition and Greater Fortune, shows only the tabs for those two figures. This suggests the possibility that a simple method for concentrating on a question and then selecting the appropriate figure would simply be to place a finger on the row of tabs. This would, of course, not produce a full geomantic tableau, but would still be in accord with the view that a channel would be opened between the questioner and the celestial spheres, which would guide the hand making the selection. It is reinforced by the fact that the table of contents is placed nearby, at folio 90r.

If this suggestion has any force, then it might follow that the user of the book would be intrigued by the response to troubling questions and attracted by the guidance and insights offered, and

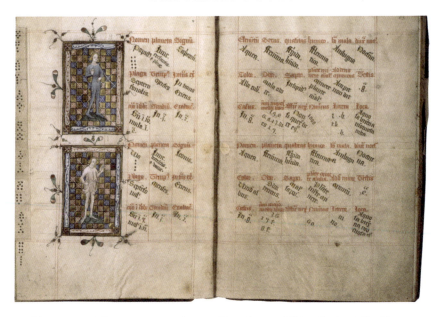

Figure 9.6. The geomantic figures Populus and Via. Oxford, Bodleian Library, MS Bodley 581, ff22v–23r.

would proceed to the more detailed parts of the book. At that point, a more traditional means of casting the points and figures would be needed – but perhaps the intention was that the giver of the manuscript would be able to help and would thus gain access to the king. The view that the manuscript, if approached in this way, was calculated to appeal directly to the king, is supported by the full title of the *Rosary*. This is *Rosary of King Richard, about the figures supporting and hindering the fulfilment of the questioner and the question, with certain information concerning the planetary hour under which questions should be asked.*

Despite the attractiveness of this manuscript, it is, as has been said, clear that it was little used. Whether this was because King Richard was not interested in geomancy specifically, or whether he preferred more personal forms of consultation, cannot be known from this evidence. However, the facts that he apparently expressed an interest in the subject and that possession of this manuscript would put the king of England on a par with rulers of other European states help to demonstrate the power and popularity of geomancy in the late middle ages and the early Renaissance. The emphasis on astrological meanings and interpretations also demonstrates that, when it came to rulers seeking information concerning political or personal developments, and guidance in making decisions, little had truly changed since the ninth century. Kings, emperors, queens and aristocrats may not have wished to practise magic themselves (with a few possible exceptions, like Alfonso X of Castile and Charles V of France). However, they were consistently interested in gaining advantageous and secret knowledge, including knowledge derived from the powers and influences of the stars and planets. Such sources of information also had the advantage that the blurred boundaries between licit and illicit queries protected the powerful against open breach of church law. Equally, their ancient origins and wealth of cultural and scientific associations made them capable of being transformed into art objects, in the form of manuscripts of great beauty and attractiveness, confirming the status, connoisseurship and superior knowledge of their possessors.

CONCLUSION

It would be tedious to repeat here the arguments set out in the introduction. The preceding chapters, it is hoped, have presented case studies of the changing fashions in medieval magic over some six centuries, with particular emphasis on the forms of magic taken seriously by intellectuals and rulers. Each case study has been based on surviving manuscripts, focusing on books which present important texts and images of magic as they were consulted and used in specific times and places. Importantly, these manuscripts are also representative of the book arts as practised in the societies which produced them, a fact which demonstrates their prominent place in cultural and social activity. Moreover, many of them are connected to powerful patrons, whose interest and belief in magic, as shown by their books, may come as a surprise to modern readers.

Significantly, not all of these texts and images explicitly identify themselves as works of magic. The dangers inherent in doing that are a reminder of the paradoxical status of magic, which in turn raises the problem of attempting to define magic. An expert on the subject has declared that defining magic is a maddening task.[1] It is certain that the problem occupied the time of many theologians across the medieval period, and their views are discussed in the relevant chapters. It is safe also to say that, while all agreed on the harmfulness of magic, and especially on the dangers of demonic intervention and deception, both their analyses and their classifications varied across the long period considered here. Some were prepared to argue for the existence, and even the value, of naturally occurring occult or magical qualities that could be harnessed by humans. Others rejected even tricks and illusions as harmful, probably demonic and potentially heretical. That variability – real, despite the authority of each individual statement – opened up a debatable area as to whether any specific text or ritual was or was not magic in a negative sense. At the same time, the use of obscure terms and exotic details in core texts of theology and canon law (caused often by their reference to cases and decisions from earlier centuries) created further ambiguity, at least in ecclesiastical courts. These factors left open a potentially large border zone between what actually happened in medieval universities and royal courts, on the one hand, and what a theoretical text said, on the other. As the chapters of this book have shown, many forms of enquiry into future events and secret knowledge could inhabit that border.

On a more pragmatic level, it was also the case that secular law tended to focus on secular crimes, such as fraud, actual or attempted harm to individuals and activities likely to threaten the power or health of the ruler. When the ruler was also the user of magic, either directly or via specialists, then such issues were unlikely to arise. Nevertheless, as the cases discussed in several chapters of this book have shown, rulers were expected to take seriously their duty of care for the spiritual welfare of their subjects, as well as to protect themselves from magical attack. Both for that reason and because of the pressures on their time, it was almost unheard of for rulers to practise magic openly themselves. That did not mean, though, that they could not gain sometimes extensive knowledge of favoured forms of magic; and several medieval kings and emperors have emerged as the owners of impressive collections of books on aspects of magic and divination.

The most prominent were Alfonso X of Castile and Charles V of France; but others had collections and volumes almost as surprising. Sadly, the catalogue of the famous library of Matthias Corvinus, king of Hungary from 1458 to 1490, does not survive. However, considerable effort has been put into tracking down individual volumes from the collection. Given the king's well-documented enthusiasm for the astrology, and the guidance it could provide, it is not surprising to find that the astrological volumes that can be linked to Matthias' patronage are impressively technical. They include new translations of classic works by Arab astrologers, as well as new commentaries on the work of Ptolemy. Matthias also attracted some of the leading astronomers and astrologers of fifteenth-century Europe to his court, and employed several of them in an unsuccessful attempt to acquire one of the best collections of books on these subjects – that of the famous astronomer known as Regiomontanus.[2] The fashion was enthusiastically followed by Matthias' courtiers, who are known to have possessed copies of works on palmistry and divination, as well as books of 'Secrets', such as the wide-ranging *Secretum secretorum*. The equally famous library of the Medici rulers of Florence appears to have had fewer individual books of astrology or divination, but one volume provides a wide-ranging selection of extracts from important magical works. This manuscript, dated 1494, includes material from Ptolemy, an extract from *Picatrix*, instructions on making magic rings, tools of Solomonic magic and Hermetic texts.[3] This is all the more interesting, since it seems that Lorenzo was reputed to possess at least one magic ring in which a demon was imprisoned. Bartolomeo Masai, a Florentine goldsmith, recorded in his diary that a terrible storm which marked Lorenzo's death was caused by this demon, which broke free when the ring's powerful owner died.[4]

What is perhaps equally surprising is that the luxury and beauty of many of the manuscripts discussed here suggest that magic often had an open and prestigious place in courtly and intellectual culture. That place was supported by two key factors. The first, and probably most powerful, was the value of what magic could offer to rulers, courtiers and ambitious scholars. The second was the cultural prestige conferred by the glamour of the classical and biblical worlds from whose knowledge magic was claimed to spring. Leading figures in the early modern Renaissance exemplify this. One such was Marsilio Ficino, who worked for both Lorenzo de' Medici and King Matthias. Ficino was clearly familiar with *Picatrix*, since he referred to it and used themes from it in his own works. Another was Heinrich Cornelius Agrippa, whose expertise in geomancy has already been mentioned (Chapter 9).[5] This is the context in which rulers famous for their patronage of Renaissance art also sought out potentially powerful and magical gems, of types similar to those specified in works of magic such as *Picatrix*. Lorenzo de' Medici's gem collection included at least seventy-six such engraved jewels.[6] A comparable enthusiasm was shown by Charles IV, king of Bohemia (r. 1346–78) and ruler of the Holy Roman Empire (r. 1355–78).[7] The experience of consulting and using such fascinating texts and objects, in rooms beautifully painted with images of the heavens and the groups of planets and stars to be found there, must have conferred a flattering sense of gaining insight into the secrets of creation.[8]

That argument is connected to something unique to this volume – namely its detailed analyses of specific manuscripts. The context in which this book is read will depend on the reader, and celestial ceilings cannot be guaranteed; but on its pages are numerous illustrations of key texts and images of magic. These are supported by frequent references to digitised images of the manuscripts. All the latter are freely available, usually on the websites of the libraries that house the manuscripts concerned. However, the websites can appear unhelpful, since they frequently provide only brief information, derived from catalogues, about each manuscript. An aim of this book is to intrigue and encourage readers who have not previously explored the world of digitised manuscripts to venture into it.

The detailed accounts of fascinating texts and images given in each chapter demonstrate that these manuscripts contain real works of magic, frequently providing tools and instructions to guide readers through actual rituals and consultations. As such they constitute evidence for the reality of both belief in magic and magical practice in high places in medieval society. They also make it possible to track the rise – and fall – of different forms of magic, and to encounter medieval magic in as direct a way as is now possible. The process of following the instructions and decoding the images also brings home the hard work involved in carrying out these forms of magic. Books like these, and their contents, inhabit a different world from that of the short charms and homely rituals that were widespread across medieval societies. For that very reason, they represent a challenge to the view that witchcraft constituted the dominant form of medieval magic.

It should also be said that this book makes no claim to be a practical manual for modern readers wishing to practise magic. Its focus is wholly on surviving manuscripts of medieval magic, on their characteristics as physical and artistic objects and on the evidence they provide as to forms, uses and users of magic. To study the manuscripts and their contents in detail is to gain insight into the medieval concept of the universe, and of the powers with which God imbued different parts of that universe. The cosmos and the celestial spheres that structured it emerge as realms of great beauty, as well as great importance. They were also unexpectedly receptive to human ambitions. The belief that it was possible for humans to identify, and even to make contact with, spirits and intelligences which could aid in the unlocking of these celestial powers was both alluring and potentially rewarding. These manuscripts are proof that such promises enticed not just those on the fringes of medieval society, but also those at the very centre.

A final point is that the manuscripts also suggest an apparent paradox: that magical texts are found in greater numbers and in more explicit forms as the medieval period progresses, even as condemnation of magic becomes stronger. As the growing body of historical scholarship on medieval magic has shown, there is no doubt that ecclesiastical leaders, in particular, cracked down harder on magic after the thirteenth century. Richard Kieckhefer's work on the rising numbers of trials for magical crimes and his survey of the growing reaction against magic make this very clear.[9] Of course, medieval rulers stood in little danger of being found guilty of criminal magic or witchcraft, especially in the secular courts, which existed to administer their own law codes. That immunity did not, however, fully protect aristocrats and leading churchmen, and certainly did not apply to retainers – as several chapters here demonstrate. This makes their ownership or use of books of magic a more risky enterprise, at least from the fourteenth century on; and yet the evidence of the manuscripts is that this was a risk worth taking. It also shows that the growing number of popes and kings who believed that they were the target of magical attacks and even assassination attempts were not necessarily deluded. The negative side of the magic found in these manuscripts is that they contain numerous and ingenious methods for bringing about the suffering, downfall and destruction of enemies.

Nevertheless, that is not the predominant nature of medieval magic. As this book has sought to demonstrate, magic offered not only insight into the workings of the created universe, and tools for seeking to manipulate its powers, but also fertile ground for the production of complex and fascinating works of art. It is important that both faces of this complex subject should be borne in mind if the power of medieval magic is to be fully appreciated.

*British Library web addresses have been provided in the Notes for reference, although
at the time of publication access is not possible*

INTRODUCTION

1. Chrétien de Troyes, romance of *Cligès*, opening of the poem (my translation). See also Chrétien de Troyes, *Cligès*, trans. B. Raffel, Yale University Press, London and New Haven, CT, 1997.
2. T. Stanmore, *Love Spells and Lost Treasure: Service magic in England from the later middle ages to the early modern era*, Cambridge University Press, Cambridge, 2023.
3. J.-P. Boudet, 'Magic at Court', in S. Page and C. Rider (eds), *The Routledge History of Medieval Magic*, Routledge, London, 2019, pp. 331–42.
4. Ibid., 331. See also A. Lawrence-Mathers and C. Escobar-Vargas, *Magic and Medieval Society*, Routledge, Abingdon and New York, 2014, pp. 4–12.
5. See D. Keck, *Angels and Angelology in the Middle Ages*, Oxford University Press, Oxford, 1998, Part 2, pp. 71–115. For a helpful summary, see A. García Avilés, 'Cosmology and magic: The angel of Mars in the *Libro de astromagia*', in S. Page and C. Rider (eds), *The Routledge History of Medieval Magic*, Routledge, London, 2019, pp. 523–32.
6. See for instance T. Heslop, 'The production of de luxe manuscripts and the patronage of King Cnut and Queen Emma', *Anglo-Saxon England*, 19 (1990), pp. 151–95; and D. Pratt, 'Kings and books in Anglo-Saxon England', *Anglo-Saxon England*, 43 (2014), pp. 297–377.
7. John of Salisbury, *Policraticus*, Book Two, Chapter 27 (see C. Burnett, 'The earliest chiromancy in the West', *Journal of the Warburg and Courtauld Institutes*, 50 ([1987], pp. 189–95).
8. L. Means, 'A translation of Martin of Spain's *De Geomancia*', in L. Matheson (ed.), *Popular and Practical Science of Medieval England*, Colleagues Press, East Lansing, 1994, pp. 61–121.

1. ASTRONOMY AND ASTROLOGY IN THE CAROLINGIAN EMPIRE: A NEW VIEW OF THE SKIES

1. Augustine, *De doctrina Christiana/ On Christian Doctrine*, ed. and trans R.P.H. Green, Clarendon Press, Oxford, 1995.
2. St Augustine, *On the Literal Meaning of Genesis*, Chapter 14.29. See Augustine, *On Genesis*, ed. J.E. Rotelle, trans. E Hill, New City Press, New York, 2002, p. 208.

3. For discussion, see D. Collins, 'Scholastics, stars and magi', in D. Collins (ed.), *The Sacred and the Sinister,* Penn State University Press, University Park, PA, 2019.
4. For discussion, see E. Peters, 'Superstition and magic from Augustine to Isidore of Seville', in K. Jolly, C. Raudvere and E. Peters (eds), *Athlone History of Witchcraft and Magic in Europe*, Vol. 3: *The Middle Ages*, Athlone, London, 2002, pp. 178–86.
5. For a brief discussion, see R. Beck, *A Brief History of Ancient Astrology*, Blackwell, Oxford, 2007, esp. pp. 1–9.
6. Bede, *On Times*, in On the Nature of Things *and* On Times, trans. C. Kendall and F. Wallis, Liverpool University Press, Liverpool, 2010, p. 113.
7. Bede, *On the Nature of Things*, in On the Nature of Things *and* On Times, pp. 87–88.
8. For a full analysis of one such copy, made in Metz c.830 for Bishop Drogo (one of Charlemagne's illegitimate sons), see E. Ramirez-Weaver, *A Saving Science: Capturing the heavens in Carolingian manuscripts*, Penn State University Press, University Park, PA, 2017.
9. See Aratus, *Phaenomena*, trans. A. Poochigian, Johns Hopkins University Press, Baltimore, MD, 2010. For Avienius' version, see *Aviénus: Les phénomènes d'Aratos*, trans. J. Soubiran, C.U.F., Paris, 1981.
10. A helpful survey is given by R. McKitterick, 'The Carolingian Renaissance of culture and learning', in J. Story (ed.), *Charlemagne: Empire and society*, Manchester University Press, Manchester, 2005, pp. 151–66.
11. For a brief account of *computus*, see S. McCluskey, *Astronomies and Cultures in Early Medieval Europe*, Cambridge University Press, Cambridge, 2000, pp. 80–84.
12. Ramirez-Weaver, *A Saving Science*.
13. The Harley manuscript can be seen at: https://www.bl.uk/manuscripts/FullDisplay.aspx?ref=Harley_MS_647
14. No digital version of this manuscript is publicly available. A digital version has been published by Brill at: https://primarysources.brillonline.com/browse/vossiani-latini/vlq-079-aratea
15. This argument is set out by Augustine in his treatise on *De divinatione daemonum* (*The Divination of Demons*); a translation of the text is available at Roger Pearse's website: https://www.roger-pearse.com/weblog/wp-content/uploads/2020/06/Augustine-De-divinatione-daemonum-translation-Gassman-2020.pdf
16. Anglo-Saxon ideas on this subject are discussed in Chapter 2.
17. See Augustine, *On Genesis*, pp. 215–50.
18. Isidore of Seville, *Etymologies*, trans. S.A. Barney, W.J. Lewis, J.A. Beach and O. Berghof, Cambridge University Press, Cambridge, 2006, p. 89.
19. For the argument that Isidore drew on the Germanicus translation of Aratus' work, and the later additions to it, see M. Laistner, *Thought and Letters in Western Europe AD 500 to 900*, Cornell University Press, New York, 1957, p. 124.
20. For a translation, see W. Stahl, R. Johnson and E. Burge (eds), *Martianus Capella and the Seven Liberal Arts*, Vol. 2: *The Marriage of Philology and Mercury*, Columbia University Press, New York, 1971.

21. This has been the subject of debate, but a clear discussion is given in E. Dekker, 'Carolingian planetary observations: The case of the Leiden planetary configuration', *Journal for the History of Astronomy*, 39/1 (2008), pp. 77–90.
22. Bede, *The Reckoning of Time*, trans. F. Wallis, Liverpool University Press, Liverpool, 1988, pp. 54–58.
23. On this, see B. Eastwood, *Ordering the Heavens: Roman astronomy and cosmology in the Carolingian Renaissance* (History of Science and Medicine Library, 4; Medieval and Early Modern Science, 8), Brill, Leiden, 2007.
24. See Pliny, *Natural History*, Vol. 1, trans. H. Rackham (Loeb Classical Library, revised edition), Harvard University Press, Cambridge, MA, 1949, pp. 189–91 and 211–13.
25. E. Dekker, 'The provenance of the stars in the Leiden "Aratea" picture book', *Journal of the Warburg and Courtauld Institutes*, 73 (2010), pp. 1–37.
26. See McCluskey, *Astronomies and Cultures in Early Medieval Europe*, pp. 114–16. For Cassiodorus' use of astronomical terminology, see Cassiodorus, *Institutiones*, ed. R.A.B. Mynors, Clarendon, Oxford, 1937, at II.7.2, pp. 154–55.
27. Gregory of Tours, *De cursu stellarum ratio*, in Bruno Krusch (ed.), *Monumenta Germaniae Historica: Scriptores rerum Merovingicarum*, Hahn, Hannover, 1885.
28. For identifications of Gregory's constellations and a map of their locations, see S. McCluskey, 'Gregory of Tours, monastic timekeeping, and early Christian attitudes to astronomy', *Isis*, 81/1 (1990), pp. 17–18.
29. On these additions, or '*scholia*', see A. Cameron, *Greek Mythography in the Roman World*, Oxford University Press, Oxford, 2007, p. 19.
30. See Dekker, 'The provenance of the stars in the Leiden "Aratea" picture book'.
31. Ibid.
32. For a brief discussion of the *Epitome* and its later impact, see E. Dekker, *Illustrating the Phaenomena: Celestial cartography in antiquity and the middle ages*, Oxford University Press, Oxford, 2013, pp. 1–3.
33. Descriptions are given in Charlemagne's will. See Einhard and Notker the Stammerer, in L. Thorpe (ed. and trans.), *Two Lives of Charlemagne*, Penguin, Harmondsworth and New York, 1969, p. 89.
34. The text of the letters is given in E. Dummler, (ed.), *Epistulae Karolini aevii II* (MGH Epistolarum Tomus IV), Berlin, 1895, pp. 241–45 and 249–53.
35. For the text, see H. Boretius (ed.), *Monumenta Germaniae Historica; Legum sectio II, Capitularia Regum Francorum, Tomus I*, Berlin, 1883, pp. 318–31.
36. For a full discussion, see T. McLeish and M. Garrison, 'Reversals in wartime: Alcuin and Charlemagne discuss retrograde motion', *Interfaces: A journal of medieval European literatures*, 8 (2021), pp. 14–53.
37. For a full analysis of this manuscript and the others discussed in this chapter, see D. Blume, M. Haffner and W. Metzger, *Sternbilder des Mittelalters: der gemalte Himmel zwischen Wissenschaft und Phantasie*, Vol. I: *Text und Katalog der Handschriften*, Akademie Verlag, Berlin, 2012, esp. pp. 321–26.
38. See R. Duits, 'Celestial transmissions: An iconographical classification of constellation cycles in manuscripts (eighth to fifteenth centuries)', *Scriptorium*, 59/2 (2005), pp. 147–202.
39. A full discussion of both globes and star maps is given in Dekker, *Illustrating the Phaenomena*, esp. Chapter 3.

40. Attention was devoted to improved means for observing the night sky by Pacificus, archdeacon of Verona, in the first half of the ninth century. The body observed, however, was a star, not a planet; and the aim, as with Gregory of Tours, was to improve time-keeping at night. See F. Guidetti, 'Texts and illustrations in Venice, Biblioteca Nazionale Marciana, MS Lat. VIII 22 (2760)', in F. Pontani (ed.), *Certissima signa: A Venice conference on Greek and Latin Astronomical texts* (Antichistica 13), Edizioni Ca'Foscari Digital Publishing, Venice, 2017, pp. 97–125.

41. See Eastwood, *Ordering the Heavens*, Chapter 3.

42. Ibid.

43. See E. Gee, 'Cicero's astronomy', *Classical Quarterly*, 51/2 (2001), pp. 520–36.

44. See B. Eastwood, 'The astronomy of Macrobius in Carolingian Europe: Dungal's letter of 811 to Charles the Great', *Early Medieval Europe*, 3/2 (1994), pp. 117–34; and *Ordering the Heavens*, Chapter 2.

45. See Stahl et al., *Martianus Capella and the Seven Liberal Arts*, Vol. 2: *The marriage of Philology and Mercury*, pp. 179–80.

46. For a translation, see T. Noble (ed. and trans.), *Charlemagne and Louis the Pious: Lives by Einhard, Notker, Ermoldus, Thegan, and the Astronomer*, Penn State University Press, University Park, PA, 2009.

47. For the annal, see B. Scholz and B. Rogers-Gardner (ed. and trans.), *Carolingian Chronicles*, University of Michigan Press, Ann Arbor, MI, 1970, p. 77.

48. See ibid., 86. For discussion and the fact that the movements attributed to Mercury are mistaken, see Dekker, 'Carolingian planetary observations', pp. 78–81.

49. Scholz and Rogers-Gardner, *Carolingian Chronicles*, p. 92.

50. For full discussion, see S. Ashley, 'What did Louis the Pious see in the night sky? A new interpretation of the Astronomer's account of Halley's Comet, 837', *Early Medieval Europe*, 21/1 (2013), pp. 27–49.

2. MONKS AND MAGIC IN ANGLO-SAXON ENGLAND

1. Augustine, *On the Divination of Demons*, in R. Deferrari (ed.), *Treatises on Marriage and Other Subjects* (Vol. 15 of *Writings of St Augustine*), 2nd edn, Catholic University of America Press, Washington, DC, 1969, pp. 417–42, esp. pp. 430–33.

2. Bede, *Ecclesiastical History of the English People*, ed. and trans. B. Colgrave and R. Mynors, rev. edn, Clarendon Press, Oxford, 1992, Book Four, Chapter 27.

3. The majority of the texts in the manuscript are included in B. Günzel (ed.), *Ælfwine's Prayerbook (London, British Library, Cotton Titus D.xxvi + xxvii)* (Henry Bradshaw Society 108), Boydell, London, 1993. Images are available online at: https://www.bl.uk/manuscripts/FullDisplay.aspx?index=0&ref=Cotton_MS_Titus_D_XXVII and https://www.bl.uk/manuscripts/FullDisplay.aspx?ref=Cotton_MS_Titus_D_XXVI&index=0

4. For editions and discussions of the texts in relation to comparable examples in other Anglo-Saxon manuscripts, see L. Chardonnens, *Anglo-Saxon Prognostics, 900–1100: Study and texts*, Brill, Leiden, 2007, esp. pp. 57–59. See also R. Liuzza, 'Anglo-Saxon prognostics in context: A survey and handlist of manuscripts', *Anglo-Saxon England*, 30 (2001), pp. 181–230.

5. See Ælfric, *De temporibus anni*, ed. M. Blake, Brewer, Cambridge, 2009.
6. See Chardonnens, *Anglo-Saxon Prognostics*, pp. 174–81; and L.S. Chardonnens, 'Mantic alphabets in medieval western manuscripts and early printed books', *Modern Philology*, 110 (2013), pp. 340–66.
7. A modern English translation of the Rule is available on the Project Gutenberg website at: https://gutenberg.org/ebooks/50040
8. Augustine, *De doctrina Christiana/ On Christian Doctrine*, ed. and trans R.P.H. Green, Oxford, Clarendon Press, 1995, esp. Book Two.
9. A translation by R. Pearse is available at: https://www.roger-pearse.com/weblog/2020/06/10/augustine-de-divinatione-daemonorum-on-the-divination-of-demons-now-online-in-english/
10. Isidore of Seville, *Etymologies*, trans. S.A. Barney, W.J. Lewis, J.A. Beach and O. Berghof, Cambridge University Press, Cambridge, 2006.
11. Bede, *Ecclesiastical History of the English People*.
12. Images of the 'Book of Life' are available at: https://www.bl.uk/catalogues/illuminatedmanuscripts/record.asp?MSID=94&CollID=21&NStart=944&_ga=2.97397599.1328240631.1574766381-1542893939.15682 10187
13. See A. Meaney, 'Old English legal and penitential penalties for "heathenism"', in S. Keynes and A. Smyth (eds), *Anglo-Saxons: Studies presented to Cyril Roy Hart*, Four Courts Press, Dublin, 2006, pp. 127–58.
14. Translations of the law codes are available in D. Whitelock (ed. and trans.), *English Historical Documents*, Vol. I: *c.500–1042*, 2nd edn, Eyre Methuen and Oxford University Press, London and New York, 1979, pp. 407–76.
15. Ibid., 561.
16. Ibid., 452.
17. Ibid., 471–76.
18. See Ælfric, *Homilies of Ælfric: A supplementary collection, being twenty-one full homilies of his middle and later career for the most part not previously edited with some shorter pieces mainly passages added to the second and third series*, ed. J. Pope, 2 vols (Early English Text Society, o.s. 259 & 260), Oxford University Press, London, 1967–68.
19. Thorpe's translation is available online at: https://en.wikisource.org/wiki/The_Homilies_of_the_Anglo-Saxon_Church/VI
20. See Jesch, J., 'Cnut', in S. Echard and R. Rouse, eds., *The Encyclopedia of Medieval Literature in Britain 530–17*, Wiley-Blackwell, Oxford, 2017; M. Townend, 'Contextualizing the "*Knútsdrápur*": Skaldic praise-poetry at the court of Cnut', *Anglo-Saxon England*, 30 (2001), pp. 145–79.
21. Chardonnens, *Anglo-Saxon Prognostics*, pp. 414–16.
22. Ibid., 291–329. The version in Titus D xxvi, folios 11v–16r, is at pp. 323–28.
23. The literature on this subject is large. A helpful place to start is: I. Draelants, 'The notion of properties: A tension between *Sciencia* and *Ars* in medieval natural philosophy and magic', in S. Page and C. Rider (eds), *The Routledge History of Medieval Magic*, Routledge, London, 2019, pp. 169–86. See also C. Fanger, 'Things done wisely by a wise enchanter: Negotiating the power of words in the thirteenth century', *Esoterica*, 1 (1999), pp. 97–132.
24. *City of God*, Book Two, Chapter 25.

3. CLERICAL SIGN-READING AND THE PROBLEM OF DIVINATION

1. See especially Matthew Paris, *Le moine et le Hasard*, Bodleian library, MS Ashmole 304, ed. A. Iafrate, Classiques Garnier, Paris, 2015; 'The workshop of fortune: St Albans and the *Sortes* manuscripts', *Scriptorium*, 46 (2012), pp. 55–87; and 'Of stars and men: Matthew Paris and the illustrations of MS Ashmole 304', *Journal of the Courtauld and Warburg Institutes*, 76 (2013), pp. 139–77.

2. See C. Burnett, 'The earliest chiromancy in the West', *Journal of the Warburg and Courtauld Institutes*, 50 (1987), pp. 189–95.

3. Available at: https://mss-cat.trin.cam.ac.uk/Manuscript/R.17.1

4. For a slightly different translation of the verse, see that of T.A. Heslop, in T. Heslop, M. Gibson and R. Pfaff (eds), *The Eadwine Psalter: Text, image, and monastic culture in twelfth-century Canterbury*, Penn State University Press, University Park, PA, 1992, p. 180.

5. For a discussion of the extent of the building works, see P. Fergusson, *Canterbury Cathedral Priory in the Age of Becket*, Yale University Press, London and New Haven, CT, 2011.

6. John of Salisbury, *Policraticus*, Book Two, Chapter 27 (see Burnett, 'The earliest chiromancy in the West'). In Book One, Chapter 12, John had already added palmists and crystal-gazers (both of which groups were apparently fashionable in the royal court at the time) to his list of practitioners of more traditional sorts of forbidden magic. For a fuller discussion of John's work and its political significance, see D. Luscombe, 'John of Salisbury and courtiers' trifles', in P. Dalton and D. Luscombe (eds), *Rulership and Rebellion in the Anglo-Norman World, c1066–c1216*, Routledge, London, 2015, pp. 141–62.

7. For discussion of Adelard's background and career, see M. Gibson, 'Adelard of Bath', in C. Burnett (ed.), *Adelard of Bath: An English scientist and Arabist of the early twelfth century*, Warburg, London, 1987, pp. 7–15.

8. William of Malmesbury, *Gesta Pontificum Anglorum*, Vol. 1, ed. and trans. M. Winterbottom, Clarendon, Oxford, 2007, p. 304.

9. See F. Barlow, *The English Church 1066–1154*, Longman, London and New York, 1979, pp. 262-63; and the *Oxford Dictionary of National Biography*, 2004: https://www.oxforddnb.com/display/10.1093/ref:odnb/9780198614128.001.0001/odnb-9780198614128-e-9157

10. See William of Malmesbury, *Gesta Pontificum*, p. 306; and L. Cochrane, *Adelard of Bath: The first English scientist*, British Museum, London, 1994, pp. 3–4.

11. See Adelard of Bath, *Quaestiones Naturales*, in C. Burnett (ed. and trans.), *Adelard of Bath, Conversations with his Nephew, On the Same and the Different, Questions on Natural Science, and On Birds*, Cambridge University Press, Cambridge, 1998, p. 90.

12. John of Worcester, *The Chronicle of John of Worcester*, Vol. 3: *The Annals from 1067 to 1140, with the Gloucester Interpolations and the continuation to 1141*, ed. and trans. J. McGurk (Oxford Medieval Texts), Clarendon, Oxford, 1998, see the year 1125.

13. Burnett, 'Earliest chiromancy in the West'.

14. On this, see for instance Augustine, *City of God*, Book Ten, Chapter 32.

15. The original manuscript can be viewed at: https://mss-cat.trin.cam.ac.uk/manuscripts/uv/view.php?n=R.17.1&n=R.17.1#?c=0&m=0&s=0&cv=0&xywh=-2668%2C-218%2C8223%2C4343; see C. Burnett, 'The Eadwine Psalter and the Western tradition of the onomancy in Pseudo-Aristotle's *Secret of Secrets*', *Archives d'histoire doctrinale et littéraire du Moyen Âge*, 55 (1988), pp. 143–67.

16. Digitised images of this manuscript, with full commentary on the texts and tables, can be viewed at: https://digital.library.mcgill.ca/ms-17/

17. For lists and analyses of the various attempts to provide numerical values for Latin letters found in the *Liber Alchandrei*, see D. Juste, 'Non-transferable knowledge: Arabic and Hebrew onomancy into Latin', *Annals of Science*, 64/4 (2011), pp. 517–29. For texts included in the collection, see Juste, *Les Alchandreana primitifs*, Brill, Leiden, 2007. For new research on onomancy in later medieval contexts, see J. Edge, 'Between natural and demonic? Onomantic divination in the later middle ages', in A. Lawrence-Mathers and C. Escobar-Vargas (eds), *Medieval Perceptions of Magic, Science, and the Natural World*, Arc Humanities Press, York, 2024.

18. See H. Fulton, ' "Mirror of the gentry": Vernacular versions of the "*Secretum Secretorum*" in medieval Wales and England', in N. Kössinger and C. Wittig (eds), *Prodesse et delectare: Case studies on didactic literature in the European Middle Ages*, De Gruyter, Berlin, 2019, pp. 57–82.

19. See Fergusson, *Canterbury Cathedral Priory in the Age of Becket*, p. 19.

20. Ibid.

21. For a full analysis of the Harley Psalter and its relationship to the Utrecht Psalter, see W. Noel, *The Harley Psalter*, Cambridge University Press, Cambridge, 1995. The twelfth-century illuminations and their date are discussed on p. 140.

22. Ibid., 185.

23. See P. Zagórska, 'The patronage of the Eadwine Psalter', *Linguistica Silesiana*, 40 (2019), pp. 23–40.

24. Fergusson, *Canterbury Cathedral Priory in the Age of Becket*, pp. 19–20.

25. See D. Fleming, 'Christian Hebrew in England with and without Jewish books', *Early Middle English*, 1/1 (2019), pp. 73–82.

26. See E. de Vischer, *Reading the Rabbis: Christian Hebraism in the works of Herbert of Bosham*, Brill, Leiden and Boston, 2014.

27. For this statement and its use in twelfth-century commentaries, see P. Van den Eynde, 'Literary note on the earliest scholastic *Commentarii in Psalmos*', *Franciscan Studies*, 14/2 (1954), pp. 121–54.

28. The manuscript can be seen at: https://digital.bodleian.ox.ac.uk/objects/5885f370-ffea-4c09-9d01-a0d4399e82af/surfaces/a13655d1-fca5-4fe2-8720-392dd59b1946/

29. This would be necessary, given that his surviving literary output has been calculated at some 7,000 pages in modern editions. See B. Weiler, 'Matthew Paris in Norway', *Revue bénédictine*, 122/1 (2012), p. 153.

30. Ibid., 176. For the importance and methods of the Cahorsin bankers, see A. Bell, C. Brooks and T. Moore, 'Interest in medieval accounts: Examples from England, 1272–1340', *History*, 94 (2009), pp. 411–33.

31. See N. Vincent, *The Holy Blood: King Henry III and the Westminster blood relic*, Cambridge University Press, New York, 2001.

32. For a summary of Matthew's career, see A. Gransden, *Historical Writing in England c.550 to c.1307*, Routledge and Kegan Paul, London, 1974, pp. 356–79.
33. See Matthew Paris, *Chronica Majora*, ed. H. Luard (Rolls Series), Vol. III, London, 1876, p. 111.
34. See Matthew Paris, *The History of St Edward the King*, trans. T. Fenster and J. Wogan-Browne (French of England Translation Series, Vol. 1), ACMRS Press and Fordham University, New York, 2008. The manuscript has been digitised and can be viewed at: https://cudl.lib.cam.ac.uk/view/MS-EE-00003-00059/1
35. See D. Connolly, *The Maps of Matthew Paris: Medieval journeys through space, time and liturgy*, Brewer, Woodbridge, 2009. The map of the route to Jerusalem can be consulted at the website of the British Library: https://www.bl.uk/collection-items/matthew-paris-itinerary-map; the map of Britain can be viewed on the same website, while an interactive version is available at: https://historiacartarum.org/annotated-claudius-map/
36. On this, see D. Kim, 'Matthew Paris, visual exegesis, and apocalyptic birds in Royal MS 14 C VII', *Electronic British Library Journal*, (2014), Article 5.
37. On Matthew's handling of prophecies, see B. Weiler, 'History, prophecy and the apocalypse in the Chronicles of Matthew Paris', *English Historical Review*, 133/561 (2018), pp. 253–83. On the impact of Joachim's teachings more widely, see the papers in M. Riedl (ed.), *A Companion to Joachim of Fiore*, Brill, Leiden, 2018.
38. The literature on Joachim and his interpreters is very large, but a key work remains: M. Reeves, *The Influence of Prophecy in the Later Middle Ages*, Oxford University Press, Oxford, 1969. More recent, and giving attention to Matthew Paris' reception of Joachim's teachings, is K. Kerby-Fulton, *Books under Suspicion: Censorship and tolerance of revelatory writing in late medieval England*, University of Notre Dame Press, Notre Dame, IN, 2006.
39. On this, see G. Crupi, 'Volvelles of knowledge: Origin and development of an instrument of scientific imagination (13th–17th centuries)', *JLIS.it*, 10/2 (2019).
40. For a discussion of the full range of Masters or Authorities named and depicted in this manuscript, see A. Iafrate, 'Pythagoras' index: Denoting authorship in *Sortes* books', in S. Rapisarda and E. Niblaeus (eds), *Dialogues among Books in Medieval Western Magic and Divination* (Micrologus' Library 65), SISMEL Edizioni del Galluzzo, Florence, 2014, pp. 77–100.
41. On books of *Sortes* in the classical period, see P. van der Horst, '*Sortes*: Sacred books as instant oracles in late antiquity', in L. Rutgers et al. (eds), *The Use of Sacred Books in the Ancient World*, Peeters, Leuven, 1998, pp. 143–73.
42. On the text and the king, see C. Burnett, 'The *Sortes regis Amalrici*: An Arabic divinatory work in the Latin Kingdom of Jerusalem?', *Scripta Mediterranea*, xix–xx (1998–99), pp. 229–37. For an edition of key parts of the text, see C. Burnett, 'What is the *Experimentarius* of Bernardus Silvestris? A preliminary survey of the material', *Archives d'histoire doctrinale et littéraire du Moyen Âge* (1977), pp. 79–125.
43. Burnett, 'The *Sortes regis Amalrici*', pp. 234–36.
44. On these diagrams and their relationship to geomancy, see Burnett 'What is the *Experimentarius* of Bernardus Silvestris?'.
45. Burnett discusses the lunar mansions and their presentation in this text in both 'What is the *Experimentarius* of Bernardus Silvestris?' and 'The *Sortes regis Amalrici*'.

46. Juste, *Les Alchandreana primitifs*, p. 657.
47. M. Meerson, 'Book is a territory: A Hebrew Book of Fortune in context', *Jewish Studies Quarterly*, 13/4 (2006), pp. 393–96.
48. The text is edited and discussed in L. Chardonnens, *Anglo-Saxon Prognostics, 900–1100: Study and texts*, Brill, Leiden, 2007, pp. 476–82. See also W. Klingshirn, 'Defining the *Sortes Sanctorum*. Gibbon, Du Cange, and early Christian lot divination', *Journal of Early Christian Studies*, 10 (2002), pp. 77–130.
49. See T. Skeat, 'An early mediaeval "Book of Fate": The *Sortes XII Patriarcharum*', *Mediaeval and Renaissance Studies*, 3 (1954), pp. 41–54.
50. For a detailed list of the computistical contents, see T. Webber, *Scribes and Scholars at Salisbury Cathedral c1075–c1125*, Clarendon, Oxford, 1992, pp. 144–45. The contributions of Salisbury scribes are discussed in several places, but no comment on the *Sortes* is given.
51. Paris, *Chronica Majora*, Vol. IV, pp. 232–33; and Vol. V, pp. 284–87.
52. G. Gasper et al. (eds), *The Scientific Works of Robert Grosseteste*, Vol. 2: *Mapping the Universe*, Oxford University Press, Oxford, 2023.
53. Adam of Eynsham, *The Life of St Hugh of Lincoln*, in D. Douie and D. Farmer (eds), *Magna Vita Sancti Hugonis*, 2nd edn, Oxford University Press, Oxford, 1985, pp. 115–19.

4. ASTRAL MAGIC IN THE COURT OF ALFONSO X

1. Recent scholarship has established that its most likely author/compiler was Abū 'l-Qāsim Maslama ibn al-Qāsim Ibrāhīm al-Qurṭubī al-Zayyāt, a scholar and theologian born in Cordoba. He travelled and studied widely, not only in al-Andalus and Arabia, but also in Crete, Egypt and Syria.
2. M. Fierro, 'Bāṭīnism in al-Andalus. Maslama b. Qāsim al-Qurṭubī (d. 353/964) author of the *Rutbat al-Hakīm* and the *Ghāyāt al-Hakīm (Picatrix)*', *Studia Islamica*, 84/2 (1996), pp. 87–112.
3. C. Burnett, 'Magic in the court of Alfonso el Sabio: The Latin translation of the *Ghāyāt al-Hakīm*', in J.-P. Boudet, M. Ostorero and A. Paravicini Bagliani (eds), *De Frédéric II à Rodolphe II: Astrologie, divination et magie dans les cours (XIII–XVII siècles)*, Galluzo, Florence, 2017, pp. 37–52.
4. For Alfonso's career, see J. Escobar and S. Thacker (ed. and trans.), *Chronicle of Alfonso X* (Studies in Romance Languages 47), University of Kentucky Press, Lexington, KY, 2002.
5. See E.S. Procter, 'The scientific works of the court of Alfonso X of Castile: The king and his collaborators', *Modern Language Review*, 40 (1945), pp. 12–29. For an overview of the translations of magical works, see P. Forshaw, 'From occult ekphrasis to magical art: Transforming text into talismanic image in the scriptorium of Alfonso X', in S. Kiyanrad, C. Theiss and L. Willer (eds), *Bilde und Schrift auf 'magischen' Artefakten*, De Gruyter, Berlin, 2018, pp. 15–47.
6. Part VII, tit. 23, leyes 1–3. See: Scott, S. (trans.), *Las Siete Partidas, Vol. 1: The Medieval Church: The world of clerics and laymen (Partida I)*, ed. R. Burns, University of Pennsylvania Press, Philadelphia, PA, 2001.
7. F. Riva, 'Est Iste Liber Maximi Secreti: Alfonso X's *Liber Razielis* and the secrets of kingship', *Neophilologus*, 104 (2020), pp. 485–502. For the *Liber*

Razielis and traditions of Jewish magic, see 'Speaking with spirits in medieval magic texts', in J. Raymond (ed.), *Conversations with Angels*, Palgrave Macmillan, London, 2011, pp. 125–49; and 'Uplifting souls: The *Liber de essentia spirituum* and the *Liber Razielis*', in C. Fanger (ed.), *Invoking Angels: Theurgic ideas and practices, thirteenth to sixteenth centuries*, Penn State University Press, University Park, PA, 2012, pp. 79–112. For Alfonso as wise ruler, see also S. Page, 'Magic and the pursuit of wisdom: The familiar spirit in the *Liber Theysolius*', *La Corónica*, 1 (2007), pp. 41–70.

8. There is no printed edition of the *Liber Razielis*, although portions of early modern versions have been edited. Sophie Page gives a list of manuscripts containing versions of the text, and a transcription of the contents of the Alfonsine version as found in the fourteenth-century manuscript Vatican, Reg. Lat. 1300. These are appendices 2 and 3 of 'Uplifting souls'. The Vatican manuscript has been digitised and is available at: https://digi.vatlib.it/view/MSS_Reg.lat.1300

9. On the surviving manuscripts, see K. Kennedy, *Alfonso X of Castile-Léon: Royal patronage, self-promotion and manuscripts in thirteenth-century Spain*, Amsterdam University Press, Amsterdam, 2019.

10. See A. García-Avilés, 'Two astromagical manuscripts of Alfonso X', *Journal of the Warburg and Courtauld Institutes*, 59 (1996), pp. 14–23. On the two Madrid manuscripts, see Kennedy, *Alfonso X of Castile-Leon*.

11. This is under the heading *Aqui comiença el libro de las formas [et] de las ymagenes que son en los cielos. [et] de las uertudes [et] de las obras que salen dellas en los cuerpos que son de yuso del cielo de la luna. que mando componer de los libros de los filosofos antiguos el mucho alto [et] onrrado don ALFONSO amador de sçiencias [et] de saberes Por la gracia de dios REY de Castiella. de Toledo. de Leon. de Gallizia. de Seuilla. de Cordoua. de Murçia. de Iahen [et] del Algarbe. Et fijo del mucho onrrado REY don Fernando Et de la REYNA Donna BEATRIZ Et fue començado este libro en el anno xxv. de su regno. Et la era de Cesar: en mill [et] trezientos [et] catorze annos. Et la del nuestro sennor ihesu christo: en mill [et] dozientos [et] setaenta [et] seys annos. Et acabase [sic]: en el xxvij. anno de su regno. Et la Era de Çesar: en mill [et] trezientos [et] xvij annos. Et la del nuestro sennor ihesu christo: en mill [et] dozientos [et] setaenta [et] ix annos].*

12. For a catalogue description of the manuscript, see https://rbmecat.patrimonionacional.es/cgi-bin/koha/opac-detail.pl?biblionumber=326

13. The text of this version of the *Book of the Forms and Images* is included in J. Fernández Montana (ed.), *Lapidario del Rey D. Alfonso X. Codice original*, facsimile edn, Madrid, 1881.

14. See García-Avilés, 'Two astromagical manuscripts of Alfonso X'. For more on Charles V and his library of magical books, see Chapter 8.

15. For the Latin text, see D. Pingree (ed.), *Picatrix: The Latin version of the Ghāyat al-hakīm*, Warburg Institute, London, 1986; and for a modern English translation, see D. Attrell and D. Porreca (trans.), *Picatrix: A medieval treatise on astral magic*, Penn State University Press, University Park, PA, 2019.

16. At this point a note on terminology and translations is important. The Latin *Picatrix*, like most Latin translations of Arabic magical works by the fourteenth century, generally used the phrase *imago magicae* (magical image) to translate the term *Tilasm*. The phrase was also shortened to 'image'. *Tilasm* – itself related to the Greek *telesma* – is the origin of the English word 'talisman'.

As this word is more specific than 'image', I have used it where a complex magical object is being described. For discussion, see G. Luck, *Arcana Mundi: Magic and the occult in the Greek and Roman worlds*, Johns Hopkins University Press, Baltimore, MD, 2006, esp. p. 49; and H.D. Rutkin, 'Astrology and magic', in I.M. Resnick (ed.), *A Companion to Albert the Great: Theology, philosophy and the sciences*, Brill, Leiden, 2012, pp. 451–505.

17. My translation; see also Attrell and Porreca, *Picatrix*, p. 44.
18. It is worth noting that the first references to lunar mansions are found in Indian works of divination by c.500 BCE. See D. Pingree, 'Some of the sources of the Ghāyat al-hakīm', *Journal of the Warburg and Courtauld Institutes*, 43 (1980), p. 8.
19. Bede, *The Reckoning of Time*, trans. F. Wallis, Liverpool University Press, Liverpool, 1988, pp. 22–23.
20. These figures and images were central to the success of the text discussed in Chapter 5 below.
21. See Pingree, 'Some of the sources of the Ghāyat al-hakīm', pp. 7–8.
22. The actual origin of this work is obscure, and the information presented brings together material from a complex range of sources. For discussion, see S. Williams, *The 'Secret of Secrets': The scholarly career of a pseudo-Aristotelian text*, University of Michigan Press, Ann Arbor, MI, 2003.
23. This work is attributed to the tenth-century author, Ahmad Ibn-Ali Ibn-Wahshiyya, who claimed to have translated it into Arabic from very ancient sources. It combines information on agriculture, astrology and magic. For discussion and partial translation, see J. Hameen-Antilla, *The Last Pagans of Iraq: Ibn Wahshiyya and his Nabatean agriculture*, Brill, Leiden, 2006.
24. D. Pingree, 'Between the *Ghāya* and *Picatrix*. I: The Spanish version', *Journal of the Warburg and Courtauld Institutes*, 44 (1981), pp. 27–56.
25. E. Poulle, 'The Alfonsine Tables and Alfonso X of Castille', *Journal for the History of Astronomy*, XIX (1988). For text and discussion, see J. Chabás and B. Goldstein, *The Alfonsine Tables of Toledo: Archimedes*, Vol. 8, Kluwer, Dordrecht, Boston and London, 2003.
26. More than two hundred copies of the Latin versions of the Tables of Toledo are cited in F.S. Pedersen, *The Toledan Tables* (Historisk-filosofiske Skrifter 24), 4 vols, Det Kongelige Danske Videnskabernes Selskab, Copenhagen, 2002.
27. For analysis, see Pingree, 'Between the *Ghāya* and *Picatrix*'. See also A. d'Agostino (ed.), *Alfonso X el Sabio. Astromagia (MS Reg. Lat. 1283a)*, Liguori, Naples, 1992. The manuscript has been digitised and is available at: https://digi.vatlib.it/view/MSS_Reg.lat.1283.pt.A
28. See Pingree, *Picatrix*, xvi–xxiv.

5. A GIFT FIT FOR A KING: THE VISIONARY ASTROLO-GY OF GEORGIUS ZOTORUS ZAPARUS FENDULUS

1. A helpful summary is given in S. Gordon, 'Necromancy and the magical reputation of Michael Scot: John Rylands Library, Latin MS 105', *Bulletin of the John Rylands Library*, 92/1 (2016), pp. 73–103.
2. Fendulus calls him 'Abumazar'; most other Latin authors used 'Albumasar'; a third variant was 'Albumazar'.

3. This manuscript has been fully digitised and can be viewed online at: https://archivesetmanuscrits.bnf.fr/ark:/12148/cc66536h
4. See J.-P. Boudet, *Entre science et nigromance*, Éditions de la Sorbonne, Paris, 2006, p. 177.
5. Ibid.
6. The work of P. Hailstone has shown that by 1220 the rulers of Sicily were actively supporting crusading activity in the eastern Mediterranean. See *Recalcitrant Crusaders? The relationship between southern Italy and Sicily, crusading and the crusader states, c.1060–1198*, Routledge, London, 2021.
7. For an overview of the illuminations and a discussion of their art-historical importance, see Georgius Zotorus Zaparus Fendulus, *Liber Astrologiae*, ed. J.-P. Verdet and M.-T. Gousset, Bibliothèque nationale de France, Paris, 1989.
8. See previous chapter.
9. The fullest analysis of the Fendulus text, with a transcription of the prologue and other sections, is given by D. Blume, *Regenten des Himmels. Astrologische Bilder in Mittelalter und Renaissance*, Akademie Verlag, Berlin, 2000, esp. pp. 34ff., 204 f. and 280, n. 1. Black-and-white reproductions of images are given in plates 1–11 and 13–46.
10. This information is given in the second chapter of the *Great Introduction*. The list gave 1,022 fixed stars and the seven wandering stars or planets. Paul Kunitzsch has pointed out that Ptolemy actually listed 1,025 fixed stars since an additional three appended to Leo were also included. See P. Kunitzsch, 'Albumasariana', *Rivista de Dipartimento di Studi Asiatici* (Naples University), 62 (2002), pp. 19–28.
11. This is on folios 2r to 5v of Lat. 7330.
12. Precise references are complicated by the varying textual divisions and terminologies found in different translations, manuscript copies and modern editions of the *Great Introduction*. A complete English translation is given in Abū MaꜤShar, *The Great Introduction to Astrology by Abū MaꜤšar*, Vol. 1: *The Arabic Original and English Translation*, ed. and trans. K. Yamamoto and C. Burnett, Brill, Leiden, 2019. The full accounts of the natures of the constellations and signs are given in Part 3 of this version of the text.
13. For a brief account of the complex origins of the various systems of decans, see J. Tester, *A History of Western Astrology*, Boydell, Woodbridge, 1987, esp. pp. 19–24.
14. See Chapter 4 above.
15. See B. Bakhouche, 'Les décans dans le *Picatrix*, traité médiévale de magie astrale', *MHNH: Revista Internacional de Investigación sobre Magia y Astrología Antiguas*, 3 (2003), pp. 135–48.
16. For a brief survey of the key concepts used by medieval European astrologers, including decans and planetary houses, see C. Burnett, 'Traditions and practices in the medieval Western Christian world', in M. Heiduk, K. Herbers and H.-C. Lehner (eds), *Prognostication in the Medieval World*, Vol. 1, De Gruyter, Berlin and Boston, 2021, pp. 491–93.
17. See Z. Blazekavic, 'Indian decans, Arabic interpretations, European images', in G. Szonyi (ed.), *European Iconography East and West*, Brill, Leiden, 1996, pp. 225–35. Further information on the Indian decans is given in D. Pingree, 'The Indian iconography of the decans and horās', *Journal of the Warburg and Courtauld Institutes*, 26 (1963), pp. 223–54.

18. See L. Thorndike (ed. and trans.), *The Sphere of Sacrobosco and its Commentators*, University of Chicago Press, Chicago, IL, 1949, pp. 118–22. See also S. Sela, 'Maimonides and Māshāʾallāh on the Ninth Orb of the signs and astrology', *Aleph*, 12 (2012), pp. 101–34.

19. See Abū MaʿShar, *The Great Introduction*, p. 551.

20. Ibid.

21. As Sacrobosco explained, the concept of rising times carried several meanings. It could be applied to planets, stars and zodiac signs, and in each case followed a cyclical pattern through the year. Students of astronomy and anyone familiar with the *Great Introduction* would understand that the rising of the signs above the viewer's horizon is meant here; but Fendulus gives no explanation.

22. Abū MaʿShar, *The Great Introduction*, Part 5, Chapter 15.

23. Ibid., Part 6.

24. See D. Abulafia, *Frederick II, a Medieval Emperor*, Oxford University Press, Oxford, 1988, Chapter 7.

25. The manuscript can be viewed online at: https://www.bl.uk/manuscripts/FullDisplay.aspx?ref=Sloane_MS_3983

26. See also Chapter 7 below.

27. The manuscript has been digitised and can be viewed online at: https://www.themorgan.org/manuscript/144038

28. For the inscriptions in the manuscript and its entry in the library catalogue of the duc de Berry, see L. Delisle, *Notice sur un livre d'astrologie de Jean duc de Berri*, Techener, Paris, 1896, pp. 6–7. Brief information on the abbot's career is also given by Delisle.

29. The manuscript can be viewed online at: https://les-tres-riches-heures.chateaudechantilly.fr/

30. This manuscript can be viewed online at: https://archivesetmanuscrits.bnf.fr/ark:/12148/cc66537r

31. Thorndike, L., 'Notes on some astronomical, astrological and mathematical manuscripts of the Bibliothèque Nationale, Paris', *Journal of the Warburg and Courtauld Institutes*, 20 (1957), pp. 112–72.

32. Ibid. See also L. Thorndike, *History of Magic and Experimental Science*, Vol. III: *The Fourteenth and Fifteenth Century*, Columbia University Press, New York, 1923, p. 590.

33. Ibid., 277.

34. S. de Phares, *Le recueil des plus célèbres astrologues*, ed. J.-P. Boudet, Vol. 1, Champion, Paris, 1997, p. 330.

35. See Chapter 8 below and references there for details.

36. For discussion of his career and works, see J. Chabás and M.-M. Saby, *The Tables of 1322 by John of Lignères*, Brepols, Turnhout, 2023.

37. E. Pellegrin, *La bibliothèque des Visconti et des Sforza, ducs de Milan, au XV siècle*, CNRS, Paris, 1955.

6. THE *ARS NOTORIA*: RITUAL MAGIC FOR STUDENTS AND SCHOLARS?

1. A fundamental contribution was made by J. Véronèse, who showed that there are two main, and significantly different, versions of the work. See Julien Véronèse,

L'Ars notoria au Moyen Âge: Introduction et édition critique, SISMEL Edizioni del Galluzzo, Florence, 2007; and 'The Ars notoria in the middle ages and modern times: Diffusion and influence(s)', in S. Rapisarda and E. Niblaeus (eds), *Dialogues among Books in Medieval Western Magic and Divination* (Micrologus' Library 65), SISMEL Edizioni del Galluzzo, Florence, 2014, pp. 147–78.

2. Sophie Page has shown that the monks of St Augustine's Abbey, Canterbury, probably owned (and used) three copies of the treatise in the fifteenth century. See her *Magic in the Cloister: Pious motives, illicit interests and occult approaches to the medieval universe*, Penn State University Press, University Park, PA, 2013, pp. 112–130.

3. See F. Klaassen, 'English manuscripts of magic, 1300–1500: A preliminary survey', in C. Fanger (ed.), *Conjuring Spirits: Texts and traditions of medieval ritual magic*, Sutton, Stroud, 1998, pp. 3–31.

4. An especially significant revision, made by John, monk of Morigny, has been analysed and edited by Claire Fanger and Nicholas Watson. See C. Fanger, 'Plundering the Egyptian treasure: John the Monk's *Book of Visions* and its relation to the Ars Notoria of Solomon', in Fanger, *Conjuring Spirits*, pp. 216–49; and N. Watson, 'John the Monk's *Book of Visions of the Blessed and Undefiled Virgin Mary, Mother of God:* Two versions of a newly discovered ritual magic text', in Fanger, *Conjuring Spirits*, pp. 163–215.

5. On magical diagrams, see S. Page, 'Medieval magical figures: Between image and text', in S. Page and C. Rider (eds), *The Routledge History of Medieval Magic*, Routledge, London, 2019, pp. 432–57; for the Ars Notoria, see pp. 442–44.

6. Art historical analysis of the diagrams is at an early stage. Preliminary points were made by M. Camille, 'Visual art in two manuscripts of the *Ars Notoria*', in Fanger, *Conjuring Spirits*, pp. 110–39.

7. C. Fanger, 'The manuscripts and their users', in John of Morigny, *Liber florum celestis doctrine, Book of the Flowers of Heavenly Teaching*, ed. C. Fanger and N. Watson, Pontifical Institute of Medieval Studies, Toronto, 2015, pp. 90–120.

8. C. Fanger, 'A burnt book', in John of Morigny, *Liber florum celestis doctrine*, pp. 121–23.

9. Ibid.

10. On the latter, see B. Grévin and J. Véronèse, 'Les "caractères" magiques au Moyen Âge central (XII–XIV siècle)', *Bibliothèque de l'École des chartes*, 162/2 (2004), pp. 305–79.

11. On *computus*, see Chapter 1.

12. See F. Wallis' website 'The Calendar and the Cloister: St John's MS 17', at: https://digital.library.mcgill.ca/ms-17/folio.php?p=5v&showitem=5v_2 ComputusRelated_11Alphabets

13. See Chapter 3, Figure 3.5.

14. See G. Ferzoco, 'Notes on Hildegard's "unknown" language and writing', in B. Mayne Kienzle et al. (eds), *A Companion to Hildegard of Bingen*, Brill, Leiden and Boston, 2014, pp. 317–22.

15. For discussion, see S. Higley, *Hildegard of Bingen's Unknown Language: An edition, translation, and discussion*, Palgrave Macmillan, New York, 2007.

16. A sense of the range of Hildegard's correspondence, visionary works and writings on natural philosophy is given in Hildegard of Bingen, *Selected Writings*, ed. and trans. M. Atherton, Penguin, London, 2001.

17. Ibid., 192.
18. Julien Véronèsc, *L'Ars notoria au Moyen Âge et à l'époque moderne. Étude d'une tradition de magie théurgique (XIIe–XVIIe siècle)*, PhD thesis, Université Paris X–Nanterre, Paris, 2004, p. 65, available online at: https://shs.hal.science/tel-02975213. English translations of both the short and the longer versions of the work are given in M. Castle (ed. and trans.), *Ars Notoria: The Notory Art of Solomon: A medieval treatise on angelic magic and the art of memory*, Inner Traditions, Rochester, VT, 2023.
19. Véronèse, *L'Ars notoria au Moyen Âge et à l'époque moderne*, p. 685.
20. See ibid., 361–78 for a full outline and discussion of these rituals.
21. John of Morigny, *Liber florum celestis doctrine*, pp. 153–78.
22. This is the view of Michael Camille in 'Visual art in two manuscripts of the ars notoria', pp. 110–42.
23. Ibid.
24. Véronèse, *L'Ars notoria au Moyen Âge et à l'époque moderne*, pp. 87–100.
25. L. Chardonnens, *Anglo-Saxon Prognostics, 900–1100: Study and texts*, Brill, Leiden, 2007, pp. 181–222.
26. J. Edge, *Onomantic Divination in Late Medieval Britain*, Boydell & Brewer for York Medieval Press, Martlesham and Rochester, NY, 2024.
27. Page, *Magic in the Cloister*.
28. This manuscript has been digitised and is available at: https://archivesetmanuscrits.bnf.fr/ark:/12148/cc77378q
29. For a full analysis of the rituals, see J. Véronèse, 'Magic, theurgy and spirituality in the medieval ritual of the Ars Notoria', in C. Fanger (ed.), *Invoking Angels: Theurgic ideas and practices, thirteenth to sixteenth centuries*, Penn State University Press, University Park, PA, 2012, pp. 37–78.
30. See Bede, *The Reckoning of Time*, trans. F. Wallis, Liverpool University Press, Liverpool, 2nd edn, 2004, pp. 54–58.
31. Ibid., 87–89.
32. Véronèse, 'Magic, theurgy and spirituality', p. 15.
33. Ibid., 17.
34. This definition goes all the way back to the *Etymologies* of Isidore of Seville, an encyclopaedic work written in the seventh century, which discussed magic and 'magi' in Book Eight. See Isidore of Seville, *Etymologies*, trans. S.A. Barney, W.J. Lewis, J.A. Beach and O. Berghof, Cambridge University Press, Cambridge, 2006, p. 182.
35. See now Castle, *Ars Notoria* for translation.
36. For a full study of Dove and his book, see L. Mitchell, 'Cultural uses of magic in fifteenth-century England', unpublished PhD thesis, University of Toronto, 2011, available at: https://tspace.library.utoronto.ca/bitstream/1807/31869/1/Mitchell_Laura_T_201111_PhD_thesis.pdf
37. See Frank Klaassen, *The Transformations of Magic*, Penn State University Press, University Park, PA, 2013, pp. 161–67.
38. The translation into English was made by Robert Turner in 1657 and can be seen at: https://archive.org/details/ars_notoria

7. KING SOLOMON AND THE DEMONS: THE BOOKS OF PROFESSIONAL MAGICIANS

1. For a full discussion, see P. Torijano, *Solomon the Esoteric King: From king to magus, development of a tradition*, Brill, Leiden, 2002.
2. A valuable survey is provided by J. Véronèse, 'Solomonic magic', in S. Page and C. Rider (eds), *The Routledge History of Medieval Magic*, Routledge, London, 2019, Chapter 14.
3. Josephus, *Jewish Antiquities*, trans. R. Marcus (Loeb Classical Library), Harvard University Press, Cambridge, MA, and London, 1934, repr. 1998, Vol. 3, Book Eight, section 42, pp. 239–44.
4. For a helpful survey of early evidence for such objects, see J. Franek and D. Urbanova, 'Invoking King Solomon on Latin-inscribed objects of practical magic from late antiquity (c300–700)', *Folia Philologica*, 144 (2021), pp. 273–308.
5. For a wide-ranging discussion, see A. Iafrate, *The Long Life of Magical Objects: A study in the Solomonic tradition*, Penn State University Press, University Park, PA, 2019.
6. D. Gehr, 'Beringarius Ganellus and the *Summa Sacre Magice*: Magic as the promotion of God's kingship', in S. Page and C. Rider (eds), *The Routledge History of Medieval Magic*, Routledge, London, 2019, pp. 237–53.
7. This survives in Kassel, Landesbibliothek und Murhardsche Bibliothek der Stadt Kassel, MS 4 astron. 3, *Summa sacrae magicae*, ff. 2r–149r. The volume has been digitised and may be viewed at: https://orka.bibliothek.uni-kassel. de/viewer/image/1343812736802/1/
8. For a full account of the trial and evidence, see E. Falgairolle, *Un envoûtement en Gévaudan en l'année 1347*, Catélan, Nîmes, 1892. For discussion, see S. Baron, 'Un procès de magie en Gévaudan et ses enjeux politiques (1347)', *Cahiers de recherches médiévales et humanistes*, 33 (2017), pp. 385–417.
9. Vatican, Biblioteca Apostolica, Sezione Archivi, Notae Orange, Miscellanea, vol. for 1320.
10. The fullest account of the case, with transcriptions of sources, is given in Abbé E. Albe, *Autour de Jean XXII: Hugues Géraud évêque de Cahors; l'affaire des poisons et des envoûtements en 1317*, J. Girma/E. Privat, Cahors and Toulouse, 1904 (Extrait du *Bull. de la Soc. des Études litt., etc., du Lot*, t. XXIX.).
11. *Magica est scientia artandi spiritus malignos et benignos per nomen dei et per nomina sua, ac per nomina seculi rerum. Unde sequitur quod magica est scientia verborum, quia omne nomen est verbum, cum verbum sit omnis res que lingua profertur, si litteris scribi possit. Multe autem scientie sunt verborum, ut gramatica, logica, rectorica, magica, sed diversi mode, quia gramatica est de verbo quo ad congruitatem, et logica quo ad discernendum veritatem, et rectorica quo ad justitiam denotandum, sed magica est de verbo quo ad spiritualem substanciam coartandum.* Transcription by Joseph Peters, online at: https://www.esotericarchives. com/ganell/ssm.htm
12. For a transcription and translation of the contents list of the Kassel manuscript, see Joseph Peters' website at: https://www.esotericarchives.com/ganell/ ssm.htm

13. See G. Sofer, 'Wearing God, consecrating body parts: Berengar Ganell's *Summa sacre magice* and *Shi'ur Qomah*', *Magic, Ritual and Witchcraft*, 16/3 (2021), pp. 304–34.

14. For an English translation of this work, see S. McGregor Mathers, *The Key of Solomon the King*, Ridgway, London, 1889. A transcription of the English version in BL MS Add. 36674 is given by J. Peters on his website: https://www.esotericarchives.com/solomon/ad36674.htm. The latter also gives comparisons with Mathers' text.

15. See N. Weill-Parot (with J. Véronèse), 'Antonio da Montolmo's *De occultis et manifestis* or *Liber Intelligentiarum*: An annotated critical edition with English translation and introduction', in C. Fanger (ed.), *Invoking Angels: Theurgic ideas and practices, thirteenth to sixteenth centuries*, Penn State University Press, University Park, PA, 2012, pp. 219–93.

16. Paris, Bibliothèque nationale de France, Lat. 7337, pp. 45a–46b; Rome, Vatican Library, Vat. Lat. 4085, f. 104r–105r. N. Weill-Parot, *Les 'images astrologiques' au Moyen Age et à la Renaissance: Spéculations intellectuelles et pratiques magiques (XIIe–XVe siècle)*, Champion, Paris, 2002, pp. 897–900: '*Experimentum naturale, verum et ineffabile ad expellendum inimicum de regno, civitate, castro, villa, domo et de omni loco ut volueris ex quo istis temporibus sagax operator multa utilia elicere potest, maxime contra serpentes. Ponam ergo practicam per quam tempore quo eram parisius magister Thomas de Pizano de Bononia, tunc medicus regis Francie, expulit societates Anglicorum de dicto regno, prout mihi tunc dicebat . . . Et nota quod omnes tres practicae trahunt originem ex prima practica Thebit bencorath in de ymaginibus. Et nota quod Albertus com<m>entator in suo Speculo dixit quod praedicte ymagines sunt mere naturales sicut receptae medicinae, nec debent esse prohibite postquam physica non prohibetur.*' See also J.-P. Boudet, *Entre science et nigromance*, Éditions de la Sorbonne, Paris, 2006, pp. 403–05; and N. Weill-Parot, 'Relatos de los poderes de las imágenes mágicas entre prescripción y prueba (siglos XII–XV)', *História Revista* (Goiânia, Brazil), 22/1 (2017), pp. 19–34.

17. Otherwise known as Thābit ibn Qurra and in Latin sources as Thebit ibn Qurra and Thebit ben Cora.

18. On this, see N. Weill-Parot, 'Cecco d'Ascoli and Antonio da Montolmo: The building of a "nigromantical" cosmology and the birth of the author-magician', in S. Page and C. Rider (eds), *The Routledge History of Medieval Magic*, Routledge, London, 2019, Chapter 17.

19. Ibid.

20. S. Giralt, 'The manuscript of a medieval necromancer: Magic in Occitan and Latin in MS Vaticano, BAV, Barb. Lat. 3589', *Revue d'Histoire des Textes*, 9 (2014), pp. 221–72.

21. See chapter on *Picatrix* above.

22. The divine name is spelt in several different ways by authors and copyists, and these will be given here where relevant: specifically 'Schemhamphoras' and 'sememphoras'.

23. Giralt, 'The manuscript of a medieval necromancer', p. 232.

24. Ibid.

25. This is now Florence, Biblioteca Medicea Laurenziana, Plut. 44, cod. 35.

26. Now Rome, Vatican Library, Vat. Lat. 5333, 1542, ff. 1r–38r. On this identification, see M. Cox, 'Similar stars and strange angels: Giorgio Anselmi's astrological magic', unpublished MA dissertation, Warburg Institute, 2015, available online at: https://sas-space.sas.ac.uk/6325/. See also Weill-Parot, *Les 'images astrologiques' au Moyen Age et à la Renaissance*, pp. 622–38; and S. Page, 'A late medieval demonic invasion of the heavens', in D. Collins (ed.), *The Sacred and the Sinister*, Penn State University Press, University Park, PA, 2019, pp. 235–56.

27. The widespread uses of astrology and astral magic in the court of Milan are analysed in M. Azzolini, *The Duke and the Stars: Astrology and politics in Renaissance Milan*, Harvard University Press, Cambridge, MA, 2013. This manuscript and its context are discussed briefly on pp. 149–51. For a full analysis of the textual contents, see L. Moulinier-Brogi and J.-P. Boudet, 'Miscellanées magico-scientifiques en Italie au Quattrocento: L'exemple du manuscrit Paris, BnF, italien 1524', *Micrologus: Natura, Scienza e Societa Medievali*, XXVII (2019), pp. 399–428.

28. A black-and-white microfilm of the manuscript can be viewed online at: https://gallica.bnf.fr/ark:/12148/btv1b10036527h/f12.item. The copy of the *Clavicula Salomonis* begins at folio 178 but without its prologue.

29. Peter of Abano, *Lucidator dubitabilium astronomiae (astrologiae)*, ed. G. Federico Vescovini, Esedra, Padua, 1992, p. 117.

30. This transitional period and the evidence supplied in surviving manuscripts are valuably analysed in F. Klaassen, *The Transformations of Magic*, Penn State University Press, University Park, PA, 2013. A rather broader approach is taken by O. Davies, *Grimoires: A history of magic books*, Oxford University Press, Oxford, 2009.

31. Online at: https://www.bl.uk/manuscripts/Viewer.aspx?ref=royal_ms_17_a_xlii_fs001r

32. For important discussions of magic in early modern England, and the context for this manuscript, see F. Klaassen, 'Curious companions: Spirit conjuring and alchemy in the sixteenth century', in M. Brock, R. Raiswell and D. Winter (eds), *Knowing Demons, Knowing Spirits in the Early Modern Period*, Palgrave Macmillan, London, 2018, pp. 145–70; and F. Klaassen, *Making Magic in Elizabethan England: Two early modern vernacular books of magic*, Penn State University Press, University Park, PA, 2019.

33. The date of composition of the Latin version of this text is debated by historians, but there is agreement that the prologue is likely to reflect the attempt by Pope John XXII to stamp out the rise of magic in the early fourteenth century. See for instance R. Kieckhefer, *Magic in the Middle Ages*, 3rd edn, Cambridge University Press, Cambridge, 2022, pp. 195–200.

34. The Latin text on which this English translation is based has been edited: G. Hedegård, *Liber Iuratus Honorii: A critical edition of the Latin version of the Sworn Book of Honorius* (Studia Latina Stockholmiensia 48), Almqvist & Wiksell International, Stockholm, 2002. The English text in Royal 17 A XLII is the basis for an online transcription and translation by J. Peterson, available at: https://www.esotericarchives.com/juratus/juratus.htm

35. For a full discussion of an engraved, metal copy of the Seal, found at Doornenburg Castle, see L. Chardonnens and J. Veenstra, 'Carved in lead and

concealed in stone: A late medieval Sigillum Dei at Doornenburg Castle', *Magic, Ritual, and Witchcraft*, 9/2 (2014), pp. 117–56.

36. Peterson has identified the main texts used; see note 33 above.

37. Identifications of plants available in England in the late middle ages are given by J. Harvey, 'The square garden of Henry the Poet', *Garden History*, 15/1 (1987), pp. 1–11.

38. See S. Page, 'Uplifting souls: The *Liber de essentia spirituum* and the *Liber Razielis*', in C. Fanger (ed.), *Invoking Angels: Theurgic ideas and practices, thirteenth to sixteenth centuries*, Penn State University Press, University Park, PA, 2012, Chapter 2.

39. See the editions by Hedegård and Peterson, cited above.

40. Klaassen, *Making Magic in Elizabethan England*, introduction.

8. CHRISTINE DE PIZAN AND ASTROLOGY IN THE FRENCH COURT

1. H. Carey, *Courting Disaster: Astrology at the English court and university in the later middle ages*, Palgrave Macmillan, New York, 1992, remains a key work in this field.

2. See C. Cannon Willard, 'Christine de Pizan the Astrologer's daughter', in Franco Simone, *Mélanges à la mémoire de Franco Simone: France et Italie dans la culture européenne*, Vol. 1, Slatkine, Paris, 1980, pp. 95–111.

3. E. Laird, 'Christine de Pizan and the controversy concerning star-study in the court of Charles V', *Culture and Cosmos*, 1/2 (1997), pp. 35–48.

4. Extracts from the text, in English, are included in R. Blumenfeld-Kosinski (ed.), *Selected Writings of Christine de Pizan*, Norton, New York and London, 1997, pp. 113–16.

5. For a modern French translation, with references, see Christine de Pizan, *Livre des faits et bonnes mœurs du sage roi Charles V*, ed. and trans. J. Blanchard and M. Quereuil, Univers Poche, Paris, 2013, pp. 223–350.

6. See L. Walters, 'The royal vernacular: Poet and patron in Christine de Pizan's *Charles V* and *Sept psaumes allégorisés*', in R. Blumenfeld-Kosinski, D. Robertson and N.B. Warren (eds), *The Vernacular Spirit*, Palgrave Macmillan, London, 2002, pp. 145–82; and L. Walters, 'Christine de Pizan as translator and voice of the body politic', in B. Altman and D. McGrady (eds), *Christine de Pizan: A casebook*, Routledge, New York, 2020.

7. Christine de Pizan, *Livre des faits et bonnes mœurs du sage roi Charles V*, Chapter 11.

8. Ibid., Chapter 12.

9. Ibid., Chapter 46.

10. Christine de Pizan, *The Book of the Body Politic*, ed. K. Langdon Forhan, Cambridge University Press, Cambridge, 1994, p. 44.

11. Ibid.

12. B. Le Cadet (ed.), *Inventaire de l'ancienne bibliothèque du Louvre*, De Bure, Paris, 1836.

13. For a preliminary survey of his main works in Latin translations, see L. Thorndike, 'The Latin translations of astrological works by Messahala', *Osiris*, 12 (1956), pp. 49–72.

14. See E. Laird and R. Fischer, *Pèlerin de Prusse on the Astrolabe*, Medieval & Renaissance Texts and Studies, Binghamton, NY, 1995, for texts and English translations of these works. Biographical information and sources are discussed on p. 1.

15. A shorter analysis of the catalogue is given in Carey, *Courting Disaster*, p. 109.

16. 'Geber' was the usual Latin version of the name of the twelfth-century, Iberian astronomer and mathematician Abū Muḥammad Jābir ibn Aflaḥ. The same name was applied to the supposed author of a large body of alchemical treatises in Latin translation, the ninth-century polymath Abū Mūsā Jābir ibn Ḥayyān. The two were frequently confused.

17. See O. Pedersen, 'The origins of the *Theorica Planetarum*', *Journal for the History of Astronomy*, 12 (1981), pp. 113–23.

18. On John and his work, see J. Chabás and B. Goldstein, 'The master and the disciple: The *Almanac* of John of Lignères and the *Ephemerides* of John of Saxony', *Journal for the History of Astronomy*, 50 (2019), pp. 82–96.

19. See J.-P. Boudet, 'The medieval Latin versions of Pseudo-Ptolemy's *Centiloquium*: A survey', in D. Juste et al. (eds), *Ptolemy's Science of the Stars in the Middle Ages*, Brepols, Turnhout, 2020, pp. 283–304.

20. Ibid., 299, and references there.

21. For a translation, see B. Dykes (ed. and trans.), *Leopold of Austria: Compilation on the science of the stars*, Cazimi Press, Minneapolis, MN, 2015.

22. The woman is also often known as the 'witch of Endor', though the biblical texts in the Book of Samuel and Chronicles focus on her possession of a spirit or talisman.

23. Text, translation and discussion are given in P. Zambelli, *The Speculum Astronomiae and its Enigma: Astrology, theology, and science in Albertus Magnus and his contemporaries*, Kluwer Academic Publishers, Dordrecht, 1992. For a more recent discussion of authorship, see J. Hackett, 'Albert the Great and the *Speculum astronomiae*: The state of the research at the beginning of the 21st century', in I. Resnick (ed.), *A Companion to Albert the Great: Theology, philosophy and the sciences*, Brill, Leiden, 2013, pp. 437–49.

24. See Chapter 3 above.

25. On the late medieval and Renaissance fashion for geomancy, see Chapter 9 below.

26. See Chapter 7 note 17 on the various spellings.

27. For discussion of the two different Latin translations of the work by Thābit ibn Qurra, see C. Burnett, 'Talismans: Magic as science? Necromancy among the seven liberal arts', in C. Burnett, *Magic and Divination in the Middle Ages*, Ashgate, Farnham, 1996.

28. F. Klaassen, *The Transformations of Magic*, Penn State University Press, University Park, PA, 2013, p. 44.

29. S. Page, *Magic in the Cloister: Pious motives, illicit interests and occult approaches to the medieval universe*, Penn State University Press, University Park, PA, 2013, pp. 17–18.

30. For an edition of a sixteenth-century English translation of this work, see M. Best and F. Brightman (eds), *The Book of Secrets of Albertus Magnus: Of the virtues of herbs, stones, and certain beasts, also a book of the marvels of the world*, Clarendon, Oxford, 1973.

31. P. Champion, *La librairie de Charles d'Orléans*, Champion, Paris, 1910, pp. 70–74.

32. E. Pellegrin, *La bibliothèque des Visconti et des Sforza, ducs de Milan, au XV siècle*, CNRS, Paris, 1955.

33. For a helpful discussion and video of the reconstruction of de Dondis' 'astrarium', see the website of the Oxford Museum of the History of Science: https://www.cabinet.ox.ac.uk/giovanni-dondis-astrarium-1364-0. See also G. Coopland, *Nicole Oresme and the Astrologers*, Liverpool University Press, Liverpool, 1952, pp. 14–16.

34. For a full discussion of the accusations and their context, see T. Adams, 'Valentina Visconti, Charles VI, and the politics of witchcraft', *Parergon*, 30/2 (2013), pp. 11–32.

35. Ibid., 40–41.

36. For text and translation, see Coopland, *Nicole Oresme and the Astrologers*, pp. 102–07.

37. For the sequence and dates of Oresme's works, see D. Di Liscia and A. Panzica, 'The writings of Nicole Oresme: A systematic inventory', *Traditio*, 77 (2022), pp. 235–375.

38. For the date of this work and the reward from the king, see L. Delisle, *Recherches sur la librairie de Charles V*, Paris, 1907, 104.

39. Adams, 'Valentina Visconti', and references there.

40. For a full discussion of these events, see J. Veenstra, *Magic and Divination at the Courts of Burgundy and France*, Brill, Leiden, 1998, esp. pp. 59–76 and 81–84.

41. For a summary of the case, see ibid., 68–69. An edition of de Bar's confession is included as appendix 1 in the same volume, pp. 341–55. See also J.-P. Boudet, 'Les condamnations de la magie à Paris en 1398', *Revue Mabillon*, 12 (2001), pp. 121–57.

42. See Steven J. Williams, *The 'Secret of Secrets': The scholarly career of a pseudo-Aristotelian text*, University of Michigan Press, Ann Arbor, MI, 2003; and C. Griffin (ed.), *The Middle English Wise Book of Philosophy and Astronomy*, parallel text edn, Winter Verlag, Heidelberg, 2013.

43. Christine de Pizan, *The Book of the Mutability of Fortune*, ed. and trans. G. Smith, Iter Press, Toronto, 2017.

44. For a translation of much of this work, see R. Blumenfeld-Kosinski, *Selected Writings of Christine de Pizan*, pp. 59–88.

45. Christine de Pizan, *Le livre du chemin de long estude*, ed. R. Püschel, Berlin, 1887, repr. Slatkine, Geneva, 1974, p. 45 (lines 1045–49).

46. Ibid., 79.

47. Ibid., 80.

48. Ibid., 84–85.

49. Christine de Pizan, *The Book of the Body Politic*, pp. 42–45 and 105–06. See also Christine de Pizan, *Le Livre du corps de policie*, ed. A. Kennedy, Champion, Paris, 1998, pp. 40–43 and 106–08.

50. Christine de Pizan, *The Book of the Body Politic*, pp. 44–45.

51. Ibid.

52. For a digital version of the Wolfenbüttel manuscript, see: https://diglib.hab.de/?db=mss&list=ms&id=18-2-aug-4f&lang=en

53. K. Broekhuijsen, 'Imagining the stars: The miniatures of the *Natuurkunde van het Geheelal* in the Herzog August Bibliothek in Wolfenbüttel', *Quaerendo*, 52/1 (2022), pp. 3–44.

54. See D. Blume, 'Children of the planets: The popularisation of astrology in the 15th century', *Micrologus: Natura, Scienza e Societa Medievali*, XII (2004), pp. 549–63; and R. Duits, 'Reading the stars of the Renaissance: Fritz Saxl and astrology', *Journal of Art Historiography*, 5 (2011), pp. 1–18.

9. DIVINATION AND POLITICS: THE RISE OF GEOMANCY

1. B. Lang, *Unlocked Books: Manuscripts of learned magic in the medieval libraries of central Europe*, Penn State University Press, University Park, PA, 2008, pp. 127–28.

2. L. Thorndike, *History of Magic and Experimental Science*, Vol. III: *The Fourteenth and Fifteenth Century*, Columbia University Press, New York, 1923, p. 119 and notes there.

3. T. Charmasson, *Recherches sur une technique divinatoire, la géomancie dans l'occident médiéval*, Droz, Paris, 1980, pp. 283–316. Another ground-breaking work, published around the same time, was S. Skinner, *Terrestrial Astrology: Divination by geomancy*, Routledge and Kegan Paul, London, 1980.

4. A. Palazzo and I. Zavattero (eds), *Geomancy and Other Forms of Divination*, SISMEL Edizioni del Galluzzo, Florence, 2017, introduction.

5. E. Savage-Smith (ed.), *Magic and Divination in Early Islam*, Ashgate, Aldershot, 2004.

6. The text of the condemnation is edited and discussed in D. Piché, *La condamnation parisienne de 1277*, Vrin, Paris, 1999; for *Estimaverunt Indi*, see p. 76.

7. A. Palazzo, 'Origin, development and decline of the western geomantic tradition', *Documenti e studi sulla tradizione filosofica medieval*, XXXI (2020), pp. 417–41, at pp. 430 and 433.

8. On Simon's poem, see most recently G. Gasper et al. (eds), *The Scientific Works of Robert Grosseteste*, Vol. 1: *Knowing and Speaking*, Oxford University Press, Oxford, 2019, pp. 31–35.

9. Thomas Aquinas, *De Sortibus; A Letter to a Friend about the Casting of Lots*, trans. P. Carey, Cascade Books, Eugene, OR, 2021.

10. On this, see L. Thorndike, *Michael Scot*, Nelson, London, 1965.

11. H. Carey, *Courting Disaster: Astrology at the English court and university in the later middle ages*, Palgrave Macmillan, New York, 1992, p. 102.

12. Ibid., 103.

13. A. de la Mare and L. Nuvoloni, *Bartolomeo Sanvito: The life and work of a Renaissance scribe*, Association internationale de bibliophilie, Paris, 2009.

14. For the geomantic work, see Carey, *Courting Disaster*, pp. 141–42. On the downfall of Duchess Eleanor and her associates, see R. Griffiths, 'The trial of Eleanor Cobham: An episode in the fall of Duke Humphrey of Gloucester', *Bulletin of the John Rylands Library*, 51/2 (1969), pp. 381–99.

15. See J. Freeman, 'Sorcery at court and manor: Margery Jourdemayne, the Witch of Eye next Westminster', *Journal of Medieval History*, 30/4 (2004), pp. 343–57.

16. L. Means, 'A translation of Martin of Spain's *De Geomancia*', in L. Matheson (ed.), *Popular and Practical Science of Medieval England*, Colleagues Press, East Lansing, 1994, pp. 61–121.

17. Ibid., 62 and 65.

18. Henry Cornelius Agrippa, *Of Geomancy*, trans. R. Turner, London, 1657, p. 7.

19. Oxford, Bodleian Library, Ashmole MS 219, folio 66r (the tableau is in the upper-right part of the page shown).

20. Pamela Benson, 'Emilia Lanier', on the website *A Critical Introduction to the Casebooks of Simon Forman and Richard Napier, 1596–1634*, https://casebooks. lib.cam.ac.uk/using-the-casebooks/meet-the-patients/emilia-lanier

21. Ibid.

22. These are discussed in K. Breen, 'A different kind of book for Richard's sake: MS Bodley 581 as ethical handbook', *The Chaucer Review*, 45/2 (2010), pp. 119–68. A detailed account of the contents of the manuscript is also given here.

CONCLUSION

1. O. Davies, *Magic: A very short introduction*, Oxford University Press, Oxford, 2012, p. 1.

2. On identifications of books, see C. Csapodi, *The Corvinian Library: History and stock*, trans. I. Gombos, rev. trans. L. András, Akadémiai Kiadó, Budapest, 1973. On the astronomers and astrologers at the court, see D. Hayton, 'Martin Bylica at the court of Matthias Corvinus: Astrology and politics in Renaissance Hungary', *Centaurus*, 49/3 (2007), pp. 185–256.

3. This is Florence, Biblioteca Medicea Laurenziana, Plut. 89, cod. sup. 38. For the full catalogue, see A. Bandini, *Catalogus Codicum Latinorum Bibliotheca Mediceae Laurentianae*, Florence, 1776 (especially Vol. III).

4. See P. Aakhus, 'Astral magic in the Renaissance: Gems, poetry and patronage of Lorenzo de Medici', *Magic, Ritual and Witchcraft*, 3/2 (2008), pp. 185–206.

5. See A. Ben-Zaken, 'Traveling with the *Picatrix*: Cultural liminalities of science and magic', in M. Fuchs et al. (eds), *Religious Individualisation: Historical dimensions and comparative perspectives*, De Gruyter, Berlin and Boston, 2020, pp. 1033–64.

6. Ibid.

7. See I. Ciulisová, 'The power of marvellous objects: Charles IV of Luxembourg, Charles V of Valois and their gemstones', *Journal of the History of Collections*, 33/1 (2021), pp. 1–13.

8. For examples of impressive astrological wall and ceiling paintings, see for instance C. Rousseau, 'Astrological imagery and rulership propaganda in the art of Cosimo I de' Medici', in N. Campion and J. Zahrt (eds), *Astrology as Art: Representation and practice*, Sophia Centre Press, University of Wales Trinity St David, Lampeter, 2018, pp. 63–85.

9. R. Kieckhefer, *European Witch Trials*, 1st edn, Routledge and Kegan Paul, London, 1976; and R. Kieckhefer, *Magic in the Middle Ages*, 3rd edn, Cambridge University Press, Cambridge, 2022, esp. Chapter 9. See also A. Lawrence-Mathers and C. Escobar-Vargas, *Magic and Medieval Society*, Routledge, Abingdon and New York, 2014, conclusion.

BIBLIOGRAPHY

MANUSCRIPTS

Basel, Universitätsbibliothek, MS AN IV 18, see: https://www.e-codices.unifr.ch/en/list/one/ubb/AN-IV-0018

Cambridge, Cambridge University Library, MS Ee 3 59, see: https://cudl.lib.cam.ac.uk/view/MS-EE-00003-00059/1

Cambridge, Trinity College, MS R 17 1 (Eadwine's Psalter), see: https://mss-cat.trin.cam.ac.uk/Manuscript/R.17.1

Chantilly, Musée Condé, MS 65 (*Très Riches Heures*), see: https://chateaude-chantilly.fr/en/collection/the-tres-riches-heures-du-duc-de-berry/

Eibingen, Abtei St Hildegard, Cod. 1. For reproductions of the miniatures, see: https://abtei-st-hildegard.de/die-scivias-miniaturen/

Kassel, Landesbibliothek und Murhardsche Bibliothek der Stadt Kassel, MS 4 astron. 3, *Summa sacrae magicae*, see: https://orka.bibliothek.uni-kassel.de/viewer/image/1343812736802/1/

Leiden, Universiteitsbibliotheek, MS VLQ 79, the Leiden *Aratea*, see: https://catalogue.leidenuniv.nl/discovery/fulldisplay?context=L&vid=31UKB_LEU:UBL_V1&search_scope=SpecialCollections&tab=SpecialCollections&docid=alma990035452500302711

London, British Library, Arundel MS 155 (the Arundel Psalter)

London, British Library, Cotton MS Claudius D VI

London, British Library, Cotton MSS Titus D xxvi and D xxvii (Aelfwine's Prayerbook)

London, British Library, Harley MS 603 (the Harley Psalter)

London, British Library, Harley MS 647

London, British Library, Harley MS 4431

London, British Library, Royal MS 14 C VII

London, British Library, Royal MS 17 A XLII

London, British Library, Sloane MS 2030

London, British Library, Sloane MS 513

London, British Library, Sloane MS 3983

London, British Library, Stowe MS 944 (New Minster 'Book of Life' or *Liber Vitae*)

London, National Art Library, MSL/1950/2464: not digitised

Madrid, Biblioteca Nacional, MS 3307, see: https://bdh-rd.bne.es/viewer.vm?id=0000122617

BIBLIOGRAPHY

Madrid, El Escorial, Real Biblioteca, MS h I 16: not digitised
New Haven, Yale University, Beinecke Rare Book and Manuscript Library, Mellon MS 1, see: https://collections.library.yale.edu/catalog/2037169
New York, Pierpont Morgan Library, MS M785, see: https://www.themorgan.org/manuscript/144038
Oxford, Bodleian Library, Ashmole MS 219, see: https://cudl.lib.cam.ac.uk/view/MS-ASHMOLE-00219/1
Oxford, Bodleian Library, Ashmole MS 226, see: https://cudl.lib.cam.ac.uk/view/MS-ASHMOLE-00224/1
Oxford, Bodleian Library, Ashmole MS 304, see: https://digital.bodleian.ox.ac.uk/objects/b6391fb2-a52e-4c69-bc13-02c04a9256a7/surfaces/50f338ff-1fa9-431a-a735-a127541af7c5/
Oxford, Bodleian Library, Ashmole MS 354: not digitised
Oxford, Bodleian Library, Ashmole MS 360, see: https://digital.bodleian.ox.ac.uk/objects/b601f070-6fcf-4c40-8b2f-31f1e79b9b34/
Oxford, Bodleian Library, Bodley MS 581, see: https://digital.bodleian.ox.ac.uk/objects/294894db-42cc-447d-a419-554fed475162/surfaces/31acdcf4-1b89-4ecb-80f8-76b2846a6388/
Oxford, Bodleian Library, MS Bodley MS 951: not digitised
Oxford, Bodleian Library, Digby MS 46, see: https://digital.bodleian.ox.ac.uk/objects/10288d68-b9fb-4707-a420-2bf157c368b0/
Oxford, Bodleian Library, Savile MS 21: not digitised
Oxford, St John's College, MS 17, see: https://digital.bodleian.ox.ac.uk/objects/66a78997-ab65-4059-a9d3-d08a0bba067c/
Paris, Bibliothèque nationale de France, MS Fr. 835, see: https://archivesetmanuscrits.bnf.fr/ark:/12148/cc779445
Paris, Bibliothèque nationale de France, MS Fr. 12779, see: https://archivesetmanuscrits.bnf.fr/ark:/12148/cc13026h
Paris, Bibliothèque nationale de France, MS Ital. 1524, see: https://gallica.bnf.fr/ark:/12148/btv1b10036527h/f12.item
Paris, Bibliothèque nationale de France, MS Lat. 7330, see: https://archivesetmanuscrits.bnf.fr/ark:/12148/cc66536h
Paris, Bibliothèque nationale de France, MS Lat. 7331, see: https://archivesetmanuscrits.bnf.fr/ark:/12148/cc66537r
Paris, Bibliothèque nationale de France, MS Lat. 9336, see: https://archivesetmanuscrits.bnf.fr/ark:/12148/cc77378q
Paris, Bibliothèque nationale de France, MS Lat. 14316, see: https://archivesetmanuscrits.bnf.fr/ark:/12148/cc750845
St Gallen, Stiftsbibliothek, MS 902, see https://www.e-codices.unifr.ch/en/list/one/csg/0902
Turin, Biblioteca Nazionale, MS E V 13: not digitised
Utrecht, Universiteitsbibliotheek, MS 32 (Utrecht Psalter), see: https://psalter.library.uu.nl/?_ga=2.178285985.482096169.1590082356-1397587295.1590082356
Vatican, Biblioteca Apostolica, Cod. Reg. Lat. 1283a, see: https://digi.vatlib.it/view/MSS_Reg.lat.1283.pt.A
Vatican, Biblioteca Apostolica, Cod. Vat. Lat. 3225 (Vatican Virgil), see: https://digi.vatlib.it/view/MSS_Vat.lat.3225

Wiesbaden, Hessische Landesbibliothek, MS 2, see: https://hlbrm.digitale-sammlungen.hebis.de/handschriften-hlbrm/content/titleinfo/449618

Wolfenbüttel, Herzog August Bibliothek, Cod. Guelf. 18.2 Aug. 4°, see: https://diglib.hab.de/?db=mss&list=ms&id=18-2-aug-4f&lang=en

PRINTED PRIMARY SOURCES

Medieval authors' names are listed under their medieval form, that is primarily as first names (e.g. Matthew) since additional descriptors (e.g. 'Paris' or 'of London') are unofficial, and usually later, additions.

Abū MaꜤShar, *The Great Introduction to Astrology by Abū MaꜤšar*, Vol. 1: *The Arabic Original and English Translation*, ed. and trans. K. Yamamoto and C. Burnett, Brill, Leiden, 2019

Adam of Eynsham, *The Life of St Hugh of Lincoln*, in D. Douie and D. Farmer (eds), *Magna Vita Sancti Hugonis*, 2nd edn, Oxford University Press, Oxford, 1985

Adelard of Bath, *Quaestiones Naturales*, in C. Burnett (ed. and trans.), *Adelard of Bath, Conversations with his Nephew, On the Same and the Different, Questions on Natural Science, and On Birds*, Cambridge University Press, Cambridge, 1998

Ælfric, *Homilies of Ælfric: A supplementary collection, being twenty-one full homilies of his middle and later career for the most part not previously edited with some shorter pieces mainly passages added to the second and third series*, ed. J. Pope, 2 vols (Early English Text Society, o.s. 259 & 260), Oxford University Press, London, 1967–68

Ælfric, *De temporibus anni*, ed. M. Blake, Brewer, Cambridge, 2009

Aratus, *Phaenomena*, trans. A. Poochigian, Johns Hopkins University Press, Baltimore, MD, 2010

Attrell, D. and D. Porreca (trans.), *Picatrix: A medieval treatise on astral magic*, Penn State University Press, University Park, PA, 2019

Augustine, *On the Divination of Demons*, in R. Deferrari (ed.), *Treatises on Marriage and Other Subjects* (Vol. 15 of *Writings of St Augustine*), 2nd edn, Catholic University of America Press, Washington, DC, 1969

Augustine, *De doctrina Christiana/On Christian Doctrine*, ed. and trans R.P.H. Green, Oxford, Clarendon Press, 1995

Augustine, *On Genesis*, ed. J.E. Rotelle, trans. E Hill, New City Press, New York, 2002

Avienius, *Aviénus: Les phénomènes d'Aratos*, trans. J. Soubiran, C.U.F., Paris, 1981

Bandini, A., *Catalogus Codicum Latinorum Bibliotheca Mediceae Laurentianae*, Florence, 1776

Bede, *Ecclesiastical History of the English People*, ed. and trans. B. Colgrave and R. Mynors, rev. edn, Clarendon Press, Oxford, 1992

Bede, *The Reckoning of Time*, trans. F. Wallis, Liverpool University Press, Liverpool, 2nd edn, 2004

Bede, On the Nature of Things *and* On Times, trans. C. Kendall and F. Wallis, Liverpool University Press, Liverpool, 2010

Best, M. and F. Brightman (eds), *The Book of Secrets of Albertus Magnus: Of the virtues of herbs, stones, and certain beasts, also a book of the marvels of the world*, Clarendon, Oxford, 1973

Cassiodorus, *Institutiones,* ed. R.A.B. Mynors, Clarendon, Oxford, 1937

Castle, M. (ed. and trans.), *Ars Notoria: The Notory Art of Solomon: A medieval treatise on angelic magic and the art of memory,* Inner Traditions, Rochester, VT, 2023

Chardonnens, L., *Anglo-Saxon Prognostics, 900–1100: Study and texts,* Brill, Leiden, 2007

Charlemagne, *Admonitio Generalis,* in H. Boretius (ed.), *Monumenta Germaniae Historica; Legum sectio II, Capitularia Regum Francorum, Tomus I,* Berlin, 1883

Christine de Pizan, *Le livre du chemin de long estude,* ed. R. Püschel, Berlin, 1887, repr. Slatkine, Geneva, 1974

Christine de Pizan, *The Book of the Body Politic,* ed. K. Langdon Forhan, Cambridge University Press, Cambridge, 1994

Christine de Pizan, *Le livre du corps de policie,* ed. A. Kennedy, Champion, Paris, 1998

Christine de Pizan, *Livre des faits et bonnes mœurs du sage roi Charles V,* ed. and trans. J. Blanchard and M. Quereuil, Univers Poche, Paris, 2013

Christine de Pizan, *The Book of the Mutability of Fortune,* ed. and trans. G. Smith, Iter Press, Toronto, 2017

d'Agostino, A. (ed.), *Alfonso X el Sabio. Astromagia (MS Reg. Lat. 1283a),* Liguori, Naples, 1992

de Phares, S., *Le recueil des plus célèbres astrologues,* ed. J.-P. Boudet, Vol. 1, Champion, Paris, 1997

Dykes, B. (ed. and trans.), *Leopold of Austria: Compilation on the science of the stars,* Cazimi Press, Minneapolis, MN, 2015

Einhard, *Life of Charlemagne,* in L. Thorpe (ed. and trans.), *Two Lives of Charlemagne,* Penguin, Harmondsworth and New York, 1969

Escobar, J. and S. Thacker (ed. and trans.), *Chronicle of Alfonso X* (Studies in Romance Languages 47), University of Kentucky Press, Lexington, KY, 2002

Fernández Montana, J. (ed.), *Lapidario del Rey D. Alfonso X. Codice original,* facsimile edn, Madrid, 1881

Georgius Zotorus Zaparus Fendulus, *Liber Astrologiae,* ed. J.-P. Verdet and M.-T. Gousset, Bibliothèque nationale de France, Paris, 1989

Gregory of Tours, *De cursu stellarum ratio,* in Bruno Krusch (ed.), *Monumenta Germaniae Historica: Scriptores rerum Merovingicarum,* Hahn, Hannover, 1885

Griffin, C. (ed.), *The Middle English Wise Book of Philosophy and Astronomy,* parallel text edn, Winter Verlag, Heidelberg, 2013

Günzel, B. (ed.), *Ælfwine's Prayerbook (London, British Library, Cotton Titus D.xxvi + xxvii)* (Henry Bradshaw Society 108), Boydell, London, 1993

Hedegård, G., *Liber Iuratus Honorii: A critical edition of the Latin version of the Sworn Book of Honorius* (Studia Latina Stockholmiensia 48), Almqvist & Wiksell International, Stockholm, 2002

Henry Cornelius Agrippa, *Of Geomancy,* trans. R. Turner, London, 1657; Wellcome Collection copy of this text may be viewed online at: https://wellcomecollection.org/works/zgvs7b2r/items

Higley, S., *Hildegard of Bingen's Unknown Language: An edition, translation, and discussion,* Palgrave Macmillan, New York, 2007

Hildegard of Bingen, *Selected Writings*, ed. and trans. M. Atherton, Penguin, London, 2001

Isidore of Seville, *Etymologies*, trans. S.A. Barney, W.J. Lewis, J.A. Beach and O. Berghof, Cambridge University Press, Cambridge, 2006

John of Morigny, *Liber florum celestis doctrine, Book of the Flowers of Heavenly Teaching*, ed. C. Fanger and N. Watson, Pontifical Institute of Medieval Studies, Toronto, 2015

John of Worcester, *The Chronicle of John of Worcester*, Vol. 3: *The Annals from 1067 to 1140, with the Gloucester Interpolations and the continuation to 1141*, ed. and trans. J. McGurk (Oxford Medieval Texts), Clarendon, Oxford, 1998

Josephus, *Jewish Antiquities*, trans. R. Marcus (Loeb Classical Library), Cambridge, MA, and London, Harvard University Press, 1934, repr. 1998

Laird, E. and R. Fischer, *Pèlerin de Prusse on the Astrolabe*, Medieval & Renaissance Texts and Studies, Binghamton, NY, 1995

Le Cadet, B. (ed.), *Inventaire de l'ancienne bibliothèque du Louvre*, De Bure, Paris, 1836

Matthew Paris, *Le moine et le Hasard* Bodleian library, MS Ashmole 304, ed. A. Iafrate, Classiques Garnier, Paris, 2015

Martianus Capella, *The Marriage of Philology and Mercury*, in W. Stahl, R. Johnson and E. Burge (eds), *Martianus Capella and the Seven Liberal Arts*, Vol. 2: *The Marriage of Philology and Mercury*, Columbia University Press, New York, 1971

Matthew Paris, *Chronica Majora*, ed. H. Luard, 6 vols (Rolls Series), London, 1872–80

Matthew Paris, *The History of St Edward the King*, trans. T. Fenster and J. Wogan-Browne (French of England Translation Series, Vol. 1), ACMRS Press and Fordham University, New York, 2008

Means, L., 'A translation of Martin of Spain's *De Geomancia*', in L. Matheson (ed.), *Popular and Practical Science of Medieval England*, Colleagues Press, East Lansing, 1994

Noble, T. (ed. and trans.), *Charlemagne and Louis the Pious: Lives by Einhard, Notker, Ermoldus, Thegan, and the Astronomer*, Penn State University Press, University Park, PA, 2009

Peter of Abano, *Lucidator dubitabilium astronomiae (astrologiae)*, ed. G. Federico Vescovini, Esedra, Padua, 1992

Pingree, D. (ed.), *Picatrix: The Latin version of the Ghāyat al-hakīm*, Warburg Institute, London, 1986

Pliny, *Natural History*, Vol. 1, trans. H. Rackham (Loeb Classical Library, revised edition), Harvard University Press, Cambridge, MA, 1949

Scholz, B. and B. Rogers-Gardner (ed. and trans.), *Carolingian Chronicles*, University of Michigan Press, Ann Arbor, MI, 1970

Scott, S. (trans.), *Las Siete Partidas*, Vol. 1: *The Medieval Church: The world of clerics and laymen (Partida I)*, ed. R. Burns, University of Pennsylvania Press, Philadelphia, PA, 2001

Thomas Aquinas, *De Sortibus; A Letter to a Friend about the Casting of Lots*, trans. P. Carey, Cascade Books, Eugene, OR, 2021

Thorndike, L. (ed. and trans.), *The Sphere of Sacrobosco and its Commentators*, University of Chicago Press, Chicago, IL, 1949

BIBLIOGRAPHY

Véronèse, J., *L'Ars notoria au Moyen Âge: Introduction et édition critique*, SISMEL Edizioni del Galluzzo, Florence, 2007

Whitelock, D. (ed. and trans.), *English Historical Documents*, Vol. I: *c. 500–1042*, 2nd edn, Eyre Methuen and Oxford University Press, London and New York, 1979

William of Malmesbury, *Gesta Pontificum Anglorum*, Vol. 1, ed. and trans. M. Winterbottom, Clarendon, Oxford, 2007

Zambelli, P., *The Speculum Astronomiae and its Enigma: Astrology, theology, and science in Albertus Magnus and his contemporaries*, Kluwer Academic Publishers, Dordrecht, 1992

SECONDARY WORKS

Aakhus, P., 'Astral magic in the Renaissance: Gems, poetry and patronage of Lorenzo de Medici', *Magic, Ritual and Witchcraft*, 3/2 (2008), pp. 185–206

Abulafia, D., *Frederick II, a Medieval Emperor*, Oxford University Press, Oxford, 1988

Adams, T., 'Valentina Visconti, Charles VI, and the politics of witchcraft', *Parergon*, 30/2 (2013), pp. 11–32

Albe, E. (Abbé), *Autour de Jean XXII: Hugues Géraud évêque de Cahors; l'affaire des poisons et des envoûtements en 1317*, J. Girma and E. Privat, Cahors and Toulouse, 1904

Ashley, S., 'What did Louis the Pious see in the night sky? A new interpretation of the Astronomer's account of Halley's Comet, 837', *Early Medieval Europe*, 21/1 (2013), pp. 27–49

Azzolini, M., *The Duke and the Stars: Astrology and politics in Renaissance Milan*, Harvard University Press, Cambridge, MA, 2013

Bakhouche, B., 'Les décans dans le *Picatrix*, traité médiévale de magie astrale', *MHNH: Revista Internacional de Investigación sobre Magia y Astrología Antiguas*, 3 (2003), pp. 135–48

Barlow, F., *The English Church 1066–1154*, Longman, London and New York, 1979

Baron, S., 'Un procès de magie en Gévaudan et ses enjeux politiques (1347)', *Cahiers de recherches médiévales et humanistes*, 33 (2017), pp. 385–417

Beck, R., *A Brief History of Ancient Astrology*, Blackwell, Oxford, 2007

Bell, A., C. Brooks and T. Moore, 'Interest in medieval accounts: Examples from England, 1272–1340', *History*, 94 (2009), pp. 411–33

Ben-Zaken, A., 'Traveling with the *Picatrix*: Cultural liminalities of science and magic', in M. Fuchs et al. (eds), *Religious Individualisation: Historical dimensions and comparative perspectives*, De Gruyter, Berlin and Boston, 2020

Blazekavic, B., 'Indian decans, Arabic interpretations, European images', in G. Szonyi (ed.), *European Iconography East and West*, Brill, Leiden, 1996

Blume, D., *Regenten des Himmels. Astrologische Bilder in Mittelalter und Renaissance*, Akademie Verlag, Berlin, 2000

Blume, D., 'Children of the planets: The popularisation of astrology in the 15th century', *Micrologus: Natura, Scienza e Societa Medievali*, XII (2004), pp. 549–63

Blume, D., M. Haffner and W. Metzger, *Sternbilder des Mittelalters: der gemalte Himmel zwischen Wissenschaft und Phantasie*, Vol. I: *Text und Katalog der Handschriften*, Akademie Verlag, Berlin, 2012

Blumenfeld-Kosinski, R. (ed.), *Selected Writings of Christine de Pizan*, Norton, New York and London, 1997

Boudet, J.-P., 'Les condamnations de la magie à Paris en 1398', *Revue Mabillon*, 12 (2001), pp. 121–57

Boudet, J.-P., *Entre science et nigromance*, Éditions de la Sorbonne, Paris, 2006

Boudet, J.-P., 'Magic at Court', in S. Page and C. Rider (eds), *The Routledge History of Medieval Magic*, Routledge, London, 2019

Boudet, J.-P., 'The medieval Latin versions of Pseudo-Ptolemy's *Centiloquium*: A survey', in D. Juste et al. (eds), *Ptolemy's Science of the Stars in the Middle Ages*, Brepols, Turnhout, 2020

Breen, K., 'A different kind of book for Richard's sake: MS Bodley 581 as ethical handbook', *The Chaucer Review*, 45/2 (2010), pp. 119–68

Broekhuijsen, K., 'Imagining the stars: The miniatures of the *Natuurkunde van het Geheelal* in the Herzog August Bibliothek in Wolfenbüttel', *Quaerendo*, 52/1 (2022), pp. 3–44

Burnett, C., 'What is the *Experimentarius* of Bernardus Silvestris? A preliminary survey of the material', *Archives d'histoire doctrinale et littéraire du Moyen Âge* (1977), pp. 79–125

Burnett, C., 'The earliest chiromancy in the West', *Journal of the Warburg and Courtauld Institutes*, 50 (1987), pp. 189–95

Burnett, C. (ed.), *Adelard of Bath: An English scientist and Arabist of the early twelfth century*, Warburg, London, 1987

Burnett, C., 'The Eadwine Psalter and the Western tradition of the onomancy in Pseudo-Aristotle's *Secret of Secrets*', *Archives d'histoire doctrinale et littéraire du Moyen Âge*, 55 (1988), pp. 143–67

Burnett, C., 'Talismans: Magic as science? Necromancy among the seven liberal arts', in C. Burnett, *Magic and Divination in the Middle Ages*, Ashgate, Farnham, 1996

Burnett, C., 'The *Sortes regis Amalrici*: An Arabic divinatory work in the Latin Kingdom of Jerusalem?', *Scripta Mediterranea*, xix–xx (1998–99), pp. 229–37

Burnett, C., 'Magic in the court of Alfonso el Sabio: The Latin translation of the *Ghāyāt al-Hakīm*', in J.-P. Boudet, M. Ostorero and A. Paravicini Bagliani (eds), *De Frédéric II à Rodolphe II: Astrologie, divination et magie dans les cours (XIIIe–XVIIe siècles)*, Galluzo, Florence, 2017

Burnett, C., 'Traditions and practices in the medieval Western Christian world', in M. Heiduk, K. Herbers and H.-C. Lehner (eds), *Prognostication in the Medieval World*, Vol. 1, De Gruyter, Berlin and Boston, 2021

Cameron, A., *Greek Mythography in the Roman World*, Oxford University Press, Oxford, 2007

Camille, M. 'Visual art in two manuscripts of the *Ars Notoria*', in C. Fanger (ed.), *Conjuring Spirits: Texts and traditions of medieval ritual magic*, Sutton, Stroud, 1998

Cannon Willard, C., 'Christine de Pizan the Astrologer's daughter', in Franco Simone, *Mélanges à la mémoire de Franco Simone: France et Italie dans la culture européenne*, Vol. 1, Slatkine, Paris, 1980

Carey, H., *Courting Disaster: Astrology at the English court and university in the later middle ages*, Palgrave Macmillan, New York, 1992

BIBLIOGRAPHY

Chabás, J. and B. Goldstein, *The Alfonsine Tables of Toledo: Archimedes*, Vol. 8, Kluwer, Dordrecht, Boston and London, 2003

Chabás, J. and B. Goldstein, 'The master and the disciple: The *Almanac* of John of Lignères and the *Ephemerides* of John of Saxony', *Journal for the History of Astronomy*, 50 (2019), pp. 82–96

Chabás, J. and M.-M. Saby, *The Tables of 1322 by John of Lignères*, Brepols, Turnhout, 2023

Champion, P., *La librairie de Charles d'Orléans*, Champion, Paris, 1910

Chardonnens, L.S., 'Mantic alphabets in medieval western manuscripts and early printed books', *Modern Philology*, 110 (2013), pp. 340–66.

Chardonnens, L. and J. Veenstra, 'Carved in lead and concealed in stone: A late medieval Sigillum Dei at Doornenburg Castle', *Magic, Ritual, and Witchcraft*, 9/2 (2014), pp. 117–56

Charmasson, T., *Recherches sur une technique divinatoire, la géomancie dans l'occident médiéval*, Droz, Paris, 1980

Ciulisová, I., 'The power of marvellous objects: Charles IV of Luxembourg, Charles V of Valois and their gemstones', *Journal of the History of Collections*, 33/1 (2021), pp. 1–13

Cochrane, L., *Adelard of Bath: The first English scientist*, British Museum, London, 1994

Collins, D., 'Scholastics, stars and magi', in D. Collins (ed.), *The Sacred and the Sinister*, Penn State University Press, University Park, PA, 2019

Connolly, D., *The Maps of Matthew Paris: Medieval journeys through space, time and liturgy*, Brewer, Woodbridge, 2009

Coopland, G., *Nicole Oresme and the Astrologers*, Liverpool University Press, Liverpool, 1952

Crupi, G., 'Volvelles of knowledge: Origin and development of an instrument of scientific imagination (13th–17th centuries)', *JLIS.it*, 10/2 (2019)

Csapodi, I., *The Corvinian Library: History and stock*, trans. I. Gombos, rev. trans. L. András, Akadémiai Kiadó, Budapest, 1973

d'Agostino, A. (ed.), *Alfonso X el Sabio: Astromagia (ms Reg. Lat. 1283a)*, Liguori, Naples, 1992

Davies, O., *Grimoires: A history of magic books*, Oxford University Press, Oxford, 2009

Davies, O., *Magic: A very short introduction*, Oxford University Press, Oxford, 2012

de Vischer, E., *Reading the Rabbis: Christian Hebraism in the works of Herbert of Bosham*, Brill, Leiden and Boston, 2014

de la Mare, A. and L. Nuvoloni, *Bartolomeo Sanvito: The life and work of a Renaissance scribe*, Association internationale de bibliophilie, Paris, 2009

Dekker, E., 'Carolingian planetary observations: The case of the Leiden planetary configuration', *Journal for the History of Astronomy*, 39/1 (2008), pp. 77–90

Dekker, E., 'The provenance of the stars in the Leiden "Aratea" picture book', *Journal of the Warburg and Courtauld Institutes*, 73 (2010), pp. 1–37

Dekker, E., *Illustrating the Phaenomena: Celestial cartography in antiquity and the middle ages*, Oxford University Press, Oxford, 2013

Delisle, L., *Notice sur un livre d'astrologie de Jean duc de Berri*, Techener, Paris, 1896

BIBLIOGRAPHY

Delisle, L., *Recherches sur la librairie de Charles V*, Paris, 1907

Di Liscia, D. and A. Panzica, 'The writings of Nicole Oresme: A systematic inventory', *Traditio*, 77 (2022), pp. 235–375

Draelants, I., 'The notion of properties: A tension between *Sciencia* and *Ars* in medieval natural philosophy and magic', in S. Page and C. Rider (eds), *The Routledge History of Medieval Magic*, Routledge, London, 2019

Duits, R., 'Celestial transmissions: An iconographical classification of constellation cycles in manuscripts (eighth to fifteenth centuries)', *Scriptorium*, 59/2 (2005), pp. 147–202

Duits, R. 'Reading the stars of the Renaissance: Fritz Saxl and astrology', *Journal of Art Historiography*, 5 (2011), pp. 1–18

Eastwood, B., 'The astronomy of Macrobius in Carolingian Europe: Dungal's letter of 811 to Charles the Great', *Early Medieval Europe*, 3/2 (1994), pp. 117–34

Eastwood, B., *Ordering the Heavens: Roman astronomy and cosmology in the Carolingian Renaissance* (History of Science and Medicine Library, 4; Medieval and Early Modern Science, 8), Brill, Leiden, 2007

Edge, J., *Onomantic Divination in Late Medieval Britain*, Boydell & Brewer for York Medieval Press, Martlesham and Rochester, NY, 2024

Edge, J. 'Between natural and demonic? Onomantic divination in the later middle ages', in A. Lawrence-Mathers and C. Escobar-Vargas (eds), *Medieval Perceptions of Magic, Science, and the Natural World*, Arc Humanities Press, York, 2024

Falgairolle, E., *Un envoûtement en Gévaudan en l'année 1347*, Catélan, Nîmes, 1892

Fanger, C. (ed.), *Conjuring Spirits: Texts and traditions of medieval ritual magic*, Sutton, Stroud, 1998

Fanger, C., 'Things done wisely by a wise enchanter: Negotiating the power of words in the thirteenth century', *Esoterica*, 1 (1999), pp. 97–132

Fanger, C. (ed.), *Invoking Angels: Theurgic ideas and practices, thirteenth to sixteenth centuries*, Penn State University Press, University Park, PA, 2012

Fergusson, P., *Canterbury Cathedral Priory in the Age of Becket*, Yale University Press, London and New Haven, CT, 2011

Ferzoco, G., 'Notes on Hildegard's "unknown" language and writing', in B. Mayne Kienzle et al. (eds), *A Companion to Hildegard of Bingen*, Brill, Leiden and Boston, 2014

Fierro, M., 'Bāṭīnism in al-Andalus. Maslama b. Qāsim al-Qurṭubī (d. 353/964) author of the *Rutbat al-Hakīm* and the *Ghāyāt al-Hakīm* (*Picatrix*)', *Studia Islamica*, 84/2 (1996), pp. 87–112

Fleming, D., 'Christian Hebrew in England with and without Jewish books', *Early Middle English*, 1/1 (2019), pp. 73–82

Forshaw, P., 'From occult ekphrasis to magical art: Transforming text into talismanic image in the scriptorium of Alfonso X', in S. Kiyanrad, C. Theiss and L. Willer (eds), *Bilde und Schrift auf 'magischen' Artefakten*, De Gruyter, Berlin, 2018

Franek, J. and D. Urbanova, 'Invoking King Solomon on Latin-inscribed objects of practical magic from late antiquity (c300–700)', *Folia Philologica*, 144 (2021), pp. 273–308

Freeman, J., 'Sorcery at court and manor: Margery Jourdemayne, the Witch of Eye next Westminster', *Journal of Medieval History*, 30/4 (2004), pp. 343–57

Fulton, H., '"Mirror of the gentry": Vernacular versions of the *"Secretum Secretorum"* in medieval Wales and England', in N. Kössinger and C. Wittig (eds), *Prodesse et delectare: Case studies on didactic literature in the European Middle Ages*, De Gruyter, Berlin, 2019

García-Avilés, A., 'Two astromagical manuscripts of Alfonso X', *Journal of the Warburg and Courtauld Institutes*, 59 (1996), pp. 14–23

García-Avilés, A., 'Cosmology and magic: The angel of Mars in the *Libro de astromagia*', in S. Page and C. Rider (eds), *The Routledge History of Medieval Magic*, Routledge, London, 2019

Gasper, G. et al. (eds), *The Scientific Works of Robert Grosseteste*, Vol. 1: *Knowing and Speaking*, Oxford University Press, Oxford, 2019

Gasper, G. et al. (eds), *The Scientific Works of Robert Grosseteste*, Vol. 2: *Mapping the Universe*, Oxford University Press, Oxford, 2023

Gee, E., 'Cicero's astronomy', *Classical Quarterly*, 51/2 (2001), pp. 520–36

Giralt, S., 'The manuscript of a medieval necromancer: Magic in Occitan and Latin in MS Vaticano, BAV, Barb. Lat. 3589', *Revue d'Histoire des Textes*, 9 (2014), pp. 221–72

Gordon, S., 'Necromancy and the magical reputation of Michael Scot: John Rylands Library, Latin MS 105', *Bulletin of the John Rylands Library*, 92/1 (2016), pp. 73–103

Gransden, A., *Historical Writing in England c. 550 to c. 1307*, Routledge and Kegan Paul, London, 1974

Grévin, B. and J. Véronèse, 'Les "caractères" magiques au Moyen Âge central (XIIᵉ–XIVᵉ siècle)', *Bibliothèque de l'École des chartes*, 162/2 (2004), pp. 305–79

Griffiths, R., 'The trial of Eleanor Cobham: An episode in the fall of Duke Humphrey of Gloucester', *Bulletin of the John Rylands Library*, 51/2 (1969), pp. 381–99

Guidetti, F., 'Texts and illustrations in Venice, Biblioteca Nazionale Marciana, MS Lat. VIII 22 (2760)', in F. Pontani (ed.), *Certissima signa: A Venice conference on Greek and Latin Astronomical texts* (Antichistica 13), Edizioni Ca'Foscari Digital Publishing, Venice, 2017

Hackett, J., 'Albert the Great and the *Speculum astronomiae*: The state of the research at the beginning of the 21st century', in I. Resnick (ed.), *A Companion to Albert the Great: Theology, philosophy and the sciences*, Brill, Leiden, 2013

Hailstone, P., *Recalcitrant Crusaders? The relationship between southern Italy and Sicily, crusading and the crusader states, c. 1060–1198*, Routledge, London, 2021

Hameen-Antilla, J., *The Last Pagans of Iraq: Ibn Wahshiyya and his Nabatean agriculture*, Brill, Leiden, 2006

Harvey, J., 'The square garden of Henry the Poet', *Garden History*, 15/1 (1987), pp. 1–11

Hayton, D., 'Martin Bylica at the court of Matthias Corvinus: Astrology and politics in Renaissance Hungary', *Centaurus*, 49/3 (2007), pp. 185–256

Heslop, T., 'The production of de luxe manuscripts and the patronage of King Cnut and Queen Emma', *Anglo-Saxon England*, 19 (1990), pp. 151–95

Heslop, T., M. Gibson and R. Pfaff (eds), *The Eadwine Psalter: Text, image, and monastic culture in twelfth-century Canterbury*, Penn State University Press, University Park, PA, 1992

Iafrate, I., 'The workshop of fortune: St Albans and the *Sortes* manuscripts', *Scriptorium*, 46 (2012), pp. 55–87

Iafrate, I., 'Of stars and men: Matthew Paris and the illustrations of MS Ashmole 304', *Journal of the Courtauld and Warburg Institutes*, 76 (2013), pp. 139–77

Iafrate, I., 'Pythagoras' index: Denoting authorship in *Sortes* books', in S. Rapisarda and E. Niblaeus (eds), *Dialogues among Books in Medieval Western Magic and Divination* (Micrologus' Library 65), SISMEL Edizioni del Galluzzo, Florence, 2014

Iafrate, I., *The Long Life of Magical Objects: A study in the Solomonic tradition*, Penn State University Press, University Park, PA, 2019

Juste, D., *Les Alchandreana primitifs*, Brill, Leiden, 2007

Juste, D., 'Non-transferable knowledge: Arabic and Hebrew onomancy into Latin', *Annals of Science*, 64/4 (2011), pp. 517–29

Keck, D. *Angels and Angelology in the Middle Ages*, Oxford University Press, Oxford, 1998

Kennedy, K., *Alfonso X of Castile-Léon: Royal patronage, self-promotion and manuscripts in thirteenth-century Spain*, Amsterdam University Press, Amsterdam, 2019

Kerby-Fulton, K., *Books under Suspicion: Censorship and tolerance of revelatory writing in late medieval England*, University of Notre Dame Press, Notre Dame, IN, 2006

Kieckhefer, R., *European Witch Trials*, 1st edn, Routledge and Kegan Paul, London, 1976

Kieckhefer, R., *Magic in the Middle Ages*, 3rd edn, Cambridge University Press, Cambridge, 2022

Kim, D., 'Matthew Paris, visual exegesis, and apocalyptic birds in Royal MS 14 C VII', *Electronic British Library Journal*, (2014), Article 5

Klaassen, F., 'English manuscripts of magic, 1300–1500: A preliminary survey', in C. Fanger (ed.), *Conjuring Spirits: Texts and traditions of medieval ritual magic*, Sutton, Stroud, 1998

Klaassen, F., *The Transformations of Magic*, Penn State University Press, University Park, PA, 2013

Klaassen, F., 'Curious companions: Spirit conjuring and alchemy in the sixteenth century', in M. Brock, R. Raiswell and D. Winter (eds), *Knowing Demons, Knowing Spirits in the Early Modern Period*, Palgrave Macmillan, London, 2018

Klaassen, F., *Making Magic in Elizabethan England: Two early modern vernacular books of magic*, Penn State University Press, University Park, PA, 2019

Klingshirn, W., 'Defining the *Sortes Sanctorum*. Gibbon, Du Cange, and early Christian lot divination', *Journal of Early Christian Studies*, 10, pp. 77–130

Kunitzsch, P., 'Albumasariana', *Rivista de Dipartimento di Studi Asiatici* (Naples University), 62 (2002), pp. 19–28

Laird, E., 'Christine de Pizan and the controversy concerning star-study in the court of Charles V', *Culture and Cosmos*, 1/2 (1997), pp. 35–48

Laistner, M., *Thought and Letters in Western Europe AD 500 to 900*, Cornell University Press, New York, 1957

Lang, B., *Unlocked Books: Manuscripts of learned magic in the medieval libraries of central Europe*, Penn State University Press, University Park, PA, 2008

Lawrence-Mathers, A. and C. Escobar-Vargas, *Magic and Medieval Society*, Routledge, Abingdon and New York, 2014

Liuzza, R., 'Anglo-Saxon prognostics in context: A survey and handlist of manuscripts', *Anglo-Saxon England*, 30 (2001), pp. 181–230

Luck, G., *Arcana Mundi: Magic and the occult in the Greek and Roman worlds*, Johns Hopkins University Press, Baltimore, MD, 2006

Luscombe, D., 'John of Salisbury and courtiers' trifles', in P. Dalton and D. Luscombe (eds), *Rulership and Rebellion in the Anglo-Norman World, c1066–c1216*, Routledge, London, 2015

McCluskey, S., 'Gregory of Tours, monastic timekeeping, and early Christian attitudes to astronomy', *Isis*, 81/1 (1990), pp. 8–22

McCluskey, S., *Astronomies and Cultures in Early Medieval Europe*, Cambridge University Press, Cambridge, 2000

McGregor Mathers, S., *The Key of Solomon the King*, Ridgway, London, 1889

McKitterick, R., 'The Carolingian Renaissance of culture and learning', in J. Story (ed.), *Charlemagne: Empire and society*, Manchester University Press, Manchester, 2005

McLeish, T. and M. Garrison, 'Reversals in wartime: Alcuin and Charlemagne discuss retrograde motion', *Interfaces: A journal of medieval European literatures*, 8 (2021), pp. 14–53

Meaney, A., 'Old English legal and penitential penalties for "heathenism"', in S. Keynes and A. Smyth (eds), *Anglo-Saxons: Studies presented to Cyril Roy Hart*, Four Courts Press, Dublin, 2006

Meerson, M., 'Book is a territory: A Hebrew Book of Fortune in context', *Jewish Studies Quarterly*, 13/4 (2006), pp. 388–411

Moulinier-Brogi, L. and J.-P. Boudet, 'Miscellanées magico-scientifiques en Italie au Quattrocento: l'exemple du manuscrit Paris, BnF, italien 1524', *Micrologus: Natura, Scienza e Societa Medievali*, XXVII (2019), pp. 399–428

Noel, W., *The Harley Psalter*, Cambridge University Press, Cambridge, 1995

Page, S., 'Magic and the pursuit of wisdom: The familiar spirit in the *Liber Theysolius*', *La Corónica*, 1 (2007), pp. 41–70

Page, S., 'Speaking with spirits in medieval magic texts', in J. Raymond (ed.), *Conversations with Angels*, Palgrave Macmillan, London, 2011

Page, S., 'Uplifting souls: The *Liber de essentia spirituum* and the *Liber Razielis*', in C. Fanger (ed.), *Invoking Angels: Theurgic ideas and practices, thirteenth to sixteenth centuries*, Penn State University Press, University Park, PA, 2012

Page, S., *Magic in the Cloister: Pious motives, illicit interests and occult approaches to the medieval universe*, Penn State University Press, University Park, PA, 2013

Page, S., 'A late medieval demonic invasion of the heavens', in D. Collins (ed.), *The Sacred and the Sinister*, Penn State University Press, University Park, PA, 2019

Page, S. and C. Rider (eds), *The Routledge History of Medieval Magic*, Routledge, London, 2019

Palazzo, A., 'Origin, development and decline of the western geomantic tradition', *Documenti e studi sulla tradizione filosofica medieval*, XXXI (2020), pp. 417–41

Palazzo, A. and I. Zavattero (eds), *Geomancy and Other Forms of Divination*, SISMEL Edizioni del Galluzzo, Florence, 2017

Pedersen, F., *The Toledan Tables* (Historisk-filosofiske Skrifter 24), 4 vols, Det Kongelige Danske Videnskabernes Selskab, Copenhagen, 2002

Pedersen, O., 'The origins of the *Theorica Planetarum*', *Journal for the History of Astronomy*, 12 (1981), pp. 113–23

Pellegrin, E., *La bibliothèque des Visconti et des Sforza, ducs de Milan, au XVe siècle*, CNRS, Paris, 1955

Peters, E., 'Superstition and magic from Augustine to Isidore of Seville', in K. Jolly, C. Raudvere and E. Peters (eds), *Athlone History of Witchcraft and Magic in Europe*, Vol. 3: *The Middle Ages*, Athlone, London, 2002

Piché, D., *La condamnation parisienne de 1277*, Vrin, Paris, 1999

Pingree, D., 'The Indian iconography of the decans and horās', *Journal of the Warburg and Courtauld Institutes*, 26 (1963), pp. 223–54

Pingree, D., 'Some of the sources of the Ghāyat al-hakīm', *Journal of the Warburg and Courtauld Institutes*, 43 (1980), pp. 1–15

Pingree, D., 'Between the *Ghāya* and *Picatrix*. I: The Spanish version', *Journal of the Warburg and Courtauld Institutes*, 44 (1981), pp. 27–56

Poulle, E., 'The Alfonsine Tables and Alfonso X of Castille', *Journal for the History of Astronomy*, XIX (1988)

Pratt, D., 'Kings and books in Anglo-Saxon England', *Anglo-Saxon England*, 43 (2014), pp. 297–377

Procter, E., 'The scientific works of the court of Alfonso X of Castile: The king and his collaborators', *Modern Language Review*, 40 (1945), pp. 12–29

Ramirez-Weaver, E., *A Saving Science: Capturing the heavens in Carolingian manuscripts*, Penn State University Press, University Park, PA, 2017

Reeves, M., *The Influence of Prophecy in the Later Middle Ages*, Oxford University Press, Oxford, 1969

Riedl, M. (ed.), *A Companion to Joachim of Fiore*, Brill, Leiden, 2018

Riva, F., 'Est Iste Liber Maximi Secreti: Alfonso X's *Liber Razielis* and the secrets of kingship', *Neophilologus*, 104 (2020), pp. 485–502

Rousseau, C., 'Astrological imagery and rulership propaganda in the art of Cosimo I de' Medici', in N. Campion and J. Zahrt (eds), *Astrology as Art: Representation and practice*, Sophia Centre Press, University of Wales Trinity St David, Lampeter, 2018

Rutkin, H., 'Astrology and magic', in I.M. Resnick (ed.), Resnick (ed.), *A Companion to Albert the Great: Theology, philosophy and the sciences*, Brill, Leiden, 2012

Savage-Smith, E. (ed.), *Magic and Divination in Early Islam*, Ashgate, Aldershot, 2004

Sela, S., 'Maimonides and Māshā'allāh on the Ninth Orb of the signs and astrology', *Aleph*, 12 (2012), pp. 101–34

Skeat, T., 'An early mediaeval "Book of Fate": The *Sortes XII Patriarcharum*', *Mediaeval and Renaissance Studies*, 3 (1954), pp. 41–54

Skinner, S., *Terrestrial Astrology: Divination by geomancy*, Routledge and Kegan Paul, London, 1980

Sofer, G., 'Wearing God, consecrating body parts: Berengar Ganell's *Summa sacre magice* and *Shi'ur Qomah*', *Magic, Ritual and Witchcraft*, 16/3 (2021), pp. 304–34

Stanmore, T., *Love Spells and Lost Treasure: Service magic in England from the late middle ages to the early modern era*, Cambridge University Press, Cambridge and New York, 2023

Tester, J., *A History of Western Astrology*, Boydell, Woodbridge, 1987

Thorndike, L., *History of Magic and Experimental Science*, Vol. III: *The Fourteenth and Fifteenth Century*, Columbia University Press, New York, 1923

Thorndike, L., 'The Latin translations of astrological works by Messahala', *Osiris*, 12 (1956), pp. 49–72

Thorndike, L. 'Notes on some astronomical, astrological and mathematical manuscripts of the Bibliothèque Nationale, Paris', *Journal of the Warburg and Courtauld Institutes*, 20 (1957), pp. 112–72

Thorndike, L., *Michael Scot*, Nelson, London, 1965

Torijano, P., *Solomon the Esoteric King: From king to magus, development of a tradition*, Brill, Leiden, 2002

Townend, M., 'Contextualizing the *"Knútsdrápur"*: Skaldic praise-poetry at the court of Cnut' *Anglo-Saxon England*, 30 (2001), pp. 145–79

Van den Eynde, P., 'Literary note on the earliest scholastic *Commentarii in Psalmos*', *Franciscan Studies*, 14/2 (1954), pp. 121–54

van der Horst, P., '*Sortes*: Sacred books as instant oracles in late antiquity', in L. Rutgers et al. (eds), *The Use of Sacred Books in the Ancient World*, Peeters, Leuven, 1998

Veenstra, J., *Magic and Divination at the Courts of Burgundy and France*, Brill, Leiden, 1998

Véronèse, J., *L'Ars notoria au Moyen Âge et à l'époque moderne. Étude d'une tradition de magie théurgique (XIIe–XVIIe siècle)*, Université Paris X–Nanterre, Paris, 2004

Véronèse, J., 'Magic, theurgy and spirituality in the medieval ritual of the Ars Notoria', in C. Fanger (ed.), *Invoking Angels: Theurgic ideas and practices, thirteenth to sixteenth centuries*, Penn State University Press, University Park, PA, 2012

Véronèse, J., 'The Ars notoria in the middle ages and modern times: Diffusion and influence(s)', in S. Rapisarda and E. Niblaeus (eds), *Dialogues among Books in Medieval Western Magic and Divination* (Micrologus' Library 65), SISMEL Edizioni del Galluzzo, Florence, 2014

Véronèse, J., 'Solomonic magic', in S. Page and C. Rider (eds), *The Routledge History of Medieval Magic*, Routledge, London, 2019

Vincent, N., *The Holy Blood: King Henry III and the Westminster blood relic*, Cambridge University Press, New York, 2001

Walters, L., 'The royal vernacular: Poet and patron in Christine de Pizan's *Charles V* and *Sept psaumes allégorisés*', in R. Blumenfeld-Kosinski, D. Robertson and N.B. Warren (eds), *The Vernacular Spirit*, Palgrave Macmillan, London, 2002

Walters, L., 'Christine de Pizan as translator and voice of the body politic', in B. Altman and D. McGrady (eds), *Christine de Pizan: A casebook*, Routledge, New York, 2020

Webber, T., *Scribes and Scholars at Salisbury Cathedral c1075–c1125*, Clarendon, Oxford, 1992

Weiler, B., 'Matthew Paris in Norway', *Revue bénédictine*, 122/1 (2012), pp. 153–81

Weiler, B., 'History, prophecy and the apocalypse in the Chronicles of Matthew Paris', *English Historical Review*, 133/561 (2018), pp. 253–83

Weill-Parot, N., *Les 'images astrologiques' au Moyen Age et à la Renaissance: Spéculations intellectuelles et pratiques magiques (XIIe–XVe siècle)*, Champion, Paris, 2002

Weill-Parot, N., 'Relatos de los poderes de las imágenes mágicas entre prescripción y prueba (siglos XII–XV)', *História Revista* (Goiânia, Brazil), 22/1 (2017), pp. 19–34

Weill-Parot, N. (with J. Véronèse), 'Antonio da Montolmo's *De occultis et manifestis* or *Liber Intelligentiarum:* An annotated critical edition with English translation and introduction', in C. Fanger (ed.), *Invoking Angels: Theurgic ideas and practices, thirteenth to sixteenth centuries*, Penn State University Press, University Park, PA, 2012

Williams, S., *The 'Secret of Secrets': The scholarly career of a pseudo-Aristotelian text*, University of Michigan Press, Ann Arbor, MI, 2003

Zagórska, P., 'The patronage of the Eadwine Psalter', *Linguistica Silesiana*, 40 (2019), pp. 23–40

ACKNOWLEDGEMENTS
AND CREDITS

This book was written across the period dominated by the coronavirus pandemic, and I should like first to thank the editors at Yale University Press, London, for their patience and support. Similar thanks go to the librarians and archivists who provided help and information even when it was not possible to visit their collections in person. I should also like to thank the University of Reading for the term of research leave which made it possible for me to make up some of the time lost, together with my colleagues in the History Department for their ongoing encouragement and interest.

On a more personal note, I thank my family for their support and interest; and give special thanks to my husband, Pete Mathers, and my sister, Helen Savigar.

INDEX

INDEX